GEOGRAPHICAL READINGS

Climatic Geomorphology

The Geographical Readings series

Published

Rivers and River Terraces G. H. DURY
Introduction to Coastline Development J. A. STEERS
Applied Coastal Geomorphology J. A. STEERS
World Vegetation Types S. R. EYRE
Developing the Underdeveloped Countries ALAN B. MOUNTJOY
Glaciers and Glacial Erosion CLIFFORD EMBLETON
Climatic Geomorphology EDWARD DERBYSHIRE
Transport and Development B. S. HOYLE

Other titles are in preparation

Climatic Geomorphology

EDITED BY
EDWARD DERBYSHIRE

MACMILLAN

First published 1973 *by*
THE MACMILLAN PRESS LTD
London and Basingstoke
Associated companies in New York Toronto
Melbourne Dublin Johannesburg and Madras

SBN 333 13652 7 (hard cover)
SBN 333 13653 5 (paper cover)

Printed in Great Britain at
THE PITMAN PRESS
Bath

Contents

ACKNOWLEDGEMENTS   7

INTRODUCTION   11

1 The Geographical Cycle   BY *William Morris Davis*   19

2 Attempt at a Classification of Climate on a Physiographic Basis   BY *Albrecht Penck*   51

3 Climate: Factor of Relief   BY *Emmanuel de Martonne*   61

4 The Cycle of Glaciation   BY *William Herbert Hobbs*   76

5 Morphology of Climatic Zones or Morphology of Landscape Belts?   BY *Siegfried Passarge*   91

6 Landforms of the Savanna Zone with a Short Dry Season   BY *Franz Thorbecke*   96

7 The Climatic Geomorphic System   BY *Julius Büdel*   104

8 The Geographic Cycle in Periglacial Regions as it is Related to Climatic Geomorphology   BY *Louis C. Peltier*   131

9 The Problem of Erosion Surfaces, Cycles of Erosion and Climatic Geomorphology   BY *Herbert Louis*   153

10 The Theory of Savanna Planation   BY *C. A. Cotton*   171

11 An Alternate Approach to Morphogenetic Climates BY *William F. Tanner*   186

12 Area Sampling for Terrain Analysis   BY *Louis C. Peltier*   193

13 Climatogenetic Geomorphology   BY *Julius Büdel*   202

14 Morphogenic Systems and Morphoclimatic Regions BY *Jean Tricart and André Cailleux*   228

15 Relationships Between Geomorphic Processes and Modern Climates as a Method in Paleoclimatology BY *Lee Wilson*   269

Index   285

Acknowledgements

The author and publishers wish to thank the following, who have kindly given permission for the use of copyright material

Chapter 1

The Geographical Cycle, by William Morris Davis, from *Geographical Journal*, **14** (1899) 481–504, by permission of the Royal Geographical Society
Glacial Erosion in France, Switzerland, and Norway, by William Morris Davis, from *Proceedings Boston Society Natural History*, **29** (1900) 273–322, by permission of Boston Society of Natural History
The Geographical Cycle in an Arid Climate, by William Morris Davis, from *Journal of Geology*, **13** (1905) 381–407

Chapter 2

Versuch einer Klimaklassification auf physiographischer Grundlage, by Albrecht Penck, translated by Roger S. Mays, from *Preussen Akademie der Wissenschaft Sitz. der physikalisch-mathematischen*, Klasse **12** (1910) 236–246, by permission of Deutsche Akademie der Wissenschaften zu Berlin

Chapter 3

Le climat-facteur du relief, by Emmanuel de Martonne, translated by Edward Derbyshire, from *Scientia* (1913) 339–355, by permission of Masson and Company

Chapter 4

The Cycle of Mountain Glaciation, by William Herbert Hobbs, from *Geographical Journal*, **35** (1910) 146–163, by permission of the Royal Geographical Society
Studies of the Cycle of Glaciation, by William Herbert Hobbs, from *Journal of Geology*, **29** (1921) 370–386

Chapter 5

Morphologie der Klimazonen oder Morphologie der Landschafts-gürtel?, by Siegfried Passarge, translated by Roger S. Mays and Edward Derbyshire, from *Petermanns Geographische Mitteilungen,* **72** (1926) 173–175, by permission of Haack Gotha

Chapter 6

Die Formenschatz im periodisch trocknen Tropenklima mit über-wiegender Regenzeit, by Franz Thorbecke, translated by Roger S. Mays and Edward Derbyshire, from *Düsseldorfer Geographische Vorträge und Erörterungen,* **3** (1927) 10–17 by permission of Ferdinand Hirt Verlag and Frau Thorbecke

Chapter 7

Das System der Klimatischen Geomorphologie, by Julius Büdel, translated by Roger S. Mays, from *Verhandl. Deutscher Geographie,* **27** (1948) 65–100, by permission of the author

Chapter 8

The Geographic Cycle in Periglacial Regions as it is related to Cli-matic Geomorphology, by Louis C. Peltier, from the *Annals of the Association of American Geographers,* **40** (1950) 214–236, by per-mission of the Association of American Geographers

Chapter 9

Rumpfflächenproblem, Erosionszyklus und Klimamorphologie, by Herbert Louis, translated by Roger S. Mays from *Geomorphologische Studien Hrsg. v. Herbert Louis u. Ing. Schaefer,* **262** (1957) 9–26

Chapter 10

The Theory of Savanna Planation, by C. A. Cotton, from *Geography,* **46** (1961) 89–101, by permission of *Geography*

Chapter 11

An Alternate Approach to Morphogenetic Climates, by William F. Tanner, from *Southeastern Geology,* **2** (1961) 251–257, by permission of *Southeastern Geology*
Geomorphology and the Sediment Transport System, by William F. Tanner, from *Southeastern Geology,* **4** (1962) 113–126, by permission of *Southeastern Geology*

Chapter 12

Area Sampling for Terrain Analysis, by Louis C. Peltier, from *The Professional Geographer*, **14** (1962), by permission of the *Association of American Geographers*

Chapter 13

Climatogenetic Geomorphology, by Julius Büdel, translated by Joyce M. Perry and Edward Derbyshire, from *Geographische Rundschau*, **15** (1963) 269–285

Chapter 14

Les grands ensembles morphoclimatiques naturels, chapters 3 and 5, from *Introduction à la Géomorphologie Climatique*, by Jean Tricart and André Cailleux, translated by Edward Derbyshire, by permission of the Longman Group Limited

Chapter 15

Les relations entre les processus géomorphologiques et le climat moderne comme méthode de paleoclimatologie, by Lee Wilson, from *Revue de géographie physique et de géologie dynamique*, **11** (1969) 303–314, by permission of Masson and Company

The publishers have made every effort to trace the copyright-holders but if they have inadvertently overlooked any, they will be pleased to make the necessary amendment at the first opportunity.

Introduction

IN company with other anthologists in this series, the editor of this volume is acutely aware of the magnitude of the task which he has set himself in attempting to represent adequately and to the satisfaction of most readers a field of knowledge at once so long established, highly contentious and fully documented as climatic geomorphology. It can only be hoped that this collection, in offering material which for linguistic or other reasons has not been widely available for critical perusal, will make possible a deeper and more balanced appreciation of the nature and development of climatic geomorphology up to the late 1960s.

The papers are arranged chronologically so that, read in sequence, they demonstrate the emergence of the main tenets of the subject and the manner in which some ideas have arisen, become submerged, and then reappeared in the search for more refined definition of the functional relationships between climate and landforms. The chosen papers may, of course, be read in isolation and studied in a comparative way. They fall into three broad groups, a reflection of the way in which the final choice was made. This was influenced by three main considerations: first, that the beginnings of the subject should be illustrated from the work of the pioneer writers; second, that the fundamental influence of the cycle concept upon the development of climatic geomorphology had to be represented; and, third, it was considered essential that examples be selected to illustrate the major climatic-geomorphological conceptual systems, notably those of Tricart and Cailleux, Büdel and Peltier.

The collection is made up of edited material from some twenty articles or chapters in books. Half of these have not previously appeared in English, so problems of translation have been added to those of editing. While some works are reproduced in essentially complete form (Passarge, Cotton, Büdel, Wilson), the remainder have been condensed to varying degrees. Wherever possible, this has been achieved by the removal of whole sections, paragraphs or passages which, in the editor's opinion, do not greatly fortify the author's main thesis. Illustrations have been edited in like manner. On the other hand, some of the original footnotes have been incorporated in the text especially when they touch upon fundamental points, as does Büdel's first footnote in his 1948 paper (page 107).

In its simplest form, the concept of a specific climate producing a distinctive suite of landforms can be traced back to Louis Agassiz and the birth of the glacial theory. However, the first generalisation and application of this concept to a range of subaerial landscapes is due to W. M. Davis. His essays on 'The Geographical Cycle' (describing humid mid-latitude, or 'normal' landscapes), 'The Glacial Cycle' and 'The Arid Cycle' laid the foundations of climatic geomorphology in the English-speaking world where it has remained a pervasive force, its influence extending, either directly or indirectly, to the work of some continental European scholars. In addition to being primordial documents in climatic geomorphology, these three papers by Davis (presented here as a single edited contribution) are important for their introduction of the cyclic model into landscape studies. Davis's model, particularly as applied to humid temperate landscapes, has been the subject of widespread and often vehement criticism in the past thirty years. It would be inappropriate to enter into this controversy here, except to suggest that Davis has been discredited largely on the basis of critical assessment of his model and its boundary conditions rather than on his manifestly vague consideration of geomorphic processes and their relative importance.

It is this deficiency which allowed him to apply the cyclic concept to glacial and arid realms, when a more rigorous consideration of process relationships would have demonstrated the futility of the attempt at that time. Thus, he applied the concept of grade, defined as a condition of balance between ability and work, to valley glaciers. This led him to suggest, for example, that glaciers wear down their valleys by abrasion and plucking to such a degree that, in time, not only are all rock steps removed, but the glacier becomes a depositional rather than an erosional agent. That is to say, the relative distribution of glacial erosion and deposition was explained not only in terms of slope but also on a temporal basis. An early development of this approach is represented by the work of W. H. Hobbs on the cycle of glaciation. Despite a more careful consideration of current work on formative processes, Hobbs explained regional contrasts in cirque morphology in temporal terms rather than in the spatial or climatic–process terms which have since been shown to offer at least as promising a basis for explanation of the variations which occur in cirques. This work proved influential especially in its perpetuation, right up to the present time, of the view that cirques are predominantly the outcome of headwall-sapping rather than landforms in which headwall retreat and basin development are broadly in equilibrium. Despite its preoccupation with the recognition and definition of cyclic stages, however, Hobb's approach to his subject is rather more circumspect, cognisant of the many unknowns involved, and less dogmatic than Davis's treatment of the glacial cycle.

Davis's essay on the arid cycle presents very general conclusions on the basis of rather specific examples. His example of an arid terrain was unduly influenced by his experience of the structurally and altitudinally distinctive Basin and Range province of the south-western United States. The model is too generalised because explanation is entirely in terms of the cyclic model, and because contemporary knowledge of the rate and relative importance of the geomorphic processes in deserts was extremely scanty at that time. Davis was led, perhaps under the influence of some of Passarge's observations, to ascribe an undue importance to wind in the development of planation surfaces. His gently-sloping plains are described as developing independently of sea-level, but the favoured process remains the cyclic one of progressive flattening of slopes. This viewpoint persisted in the English literature for many years, to some extent due to its acceptance and perpetuation by C. A. Cotton. Later work, particularly by German geomorphologists, has suggested that the interplay and denudational potential of linear erosion on the one hand and various types of mass movement on the other, varies with climate and so controls the slope assemblage in any landscape. The paper by Herbert Louis presented here is important as it develops this idea at some length, sometimes in terms highly reminiscent of, for example, Kesseli and Hack. Louis suggests that planar surfaces are formed in this way in several regions including arid ones, and as no phase of incision precedes their appearance, they differ fundamentally from Davis's peneplain. The major implications of this for the interpretation of high-level erosion remnants in extra-tropical areas have been explored by Büdel and are set out in the paper by Cotton.

Working within the tradition of Davisian cyclic geomorphology, Peltier attempted in 1950 a generalised definition of morphoclimatic regions on the basis of gross climatic parameters in a paper which is undoubtedly the best-known article on climatic geomorphology in the English language. It is reproduced here almost in its entirety, partly for its historical importance and also because it has given rise to similar inductive attempts by other American workers, notably Leopold, Wolman and Miller, Tanner and Wilson. As an initial attempt, Peltier's paper is interesting but, at the same time, it was clearly premature in that the precise relationships between geomorphic processes and climatic parameters as gross as mean annual temperature and precipitation are simply not known. Accordingly, the result is highly qualitative and rather subjective. Some variations on this approach have appeared from time to time since 1950. Tanner's work, represented here by edited sections of two of his papers, differs from Peltier's in that annual potential evaporation is substituted for mean annual temperature. The broad morphoclimatic provinces so derived are considered in terms of the directional weathering of rocks, differences in which arise from lithology and structure. The conclusion evident is that this factor is likely to militate against the establishment of clear relationships between climate and landforms. Wilson's approach is a combination of the climograph and Peltier's morphogenetic regions which are modified in the light of modern work (notably that by Langbein and Schumm on variations in sediment yield with precipitation) and renamed climate–process systems. Using a genetic classification of climate, Wilson suggests general relationships between process and landform on the one hand and air mass climatology on the other. His proposal that the technique may be of use as a paleoclimatic method is tentative: his example, drawn from South America, serves to emphasise how tenuous such arguments must remain for the present. Regional circulation models for the glacial periods would seem to rest more happily on sedimentological rather than on strictly geomorphic grounds.

The seeds of what was destined to become the distinctive continental or European school of climatic geomorphology are to be found in two early works appearing here in English for the first time. Albrecht Penck's paper, in attempting to use 'physiographic' or geomorphological criteria to subdivide climate, is in the same tradition as the work of the early ecological climatologists which culminated in the various classifications of W. Köppen. Penck's approach is essentially a hydrological one based on first principles.

It provides little information on distinctive landforms, its value being rather in its succinct statement of the general relationships controlling the movement of water in humid, nival and arid climates. Some of these are still in need of precise definition, over sixty years after the publication of the paper. Emmanuel de Martonne's article, on the other hand, is more overtly morphological, and contains much perceptive comment. For example, he singles out the forms of fluvial erosion as being particularly expressive of climatic variety and so foreshadows modern quantitative work (see, for example, the 1962 article by Peltier) which has demonstrated that climatic differences are a major source of variation in some morphometric entities. His introduction of the scale factor in distinguishing structural from climatic geomorphology was destined to be developed very much later in the work of the French school especially by Tricart, Cailleux and Birot. The brief description which he gives of landscape facies based principally on the type and degree of weathering plus the role of surface water, both controlled by climate, looks forward to later work in climatic geomorphology, while his consideration of the effect of climatic change in the partial replacement of one landform assemblage by another to produce a polycyclic landscape introduces the historical parameter and so looks forward to what came to be called climatogenetic geomorphology.

The Düsseldorf conference of 1926 on 'The Morphology of the Climatic Zones' was a milestone in the development of the subject. The published transactions contain a wealth of observation and argument, running to 100 pages, which has proved impossible to condense satisfactorily. Several important papers, including Mortensen on the Mediterranean lands, Jaeger on the seasonally wet tropics and Machatschek on inland and mountain deserts, have good claims for inclusion. However, the choice finally fell on 'Landforms of the Savanna Zone with a Short Dry Season', by Thorbecke, the convener of the conference. The reasons which dictated this choice all arise from his careful description and appraisal of the role of climatic and geomorphological *events* in the genesis of landform characteristics. His context is 'a morphologically effective dry season' making his work an early but direct precursor of an approach which is only now showing signs of quantitative development. Thorbecke draws particular attention to the transitional climatic zones as likely to illuminate the central problems of climatic geomorphology and, in his final section, provides an early description of parallel retreat of slopes and the isolation of scarp salients to form inselbergs by gully incision of sound rock followed by parallel retreat. This process is described as currently active and a product of the prevailing climate, a view which contrasts in detail with that of Büdel as may be seen in the paper by Cotton reproduced in this volume.

The validity of the approach expressed by the title of the Düsseldorf conference was questioned in a brief paper by Passarge, translated here in full. While his dismissal of any *a priori* case in favour of the description of morphological assemblages on a gross climatic basis may be left to speak for itself, it is interesting to note that in making his plea for the delimitation of landscape belts (derived from a variety of factors including climate) as sounder practice, Louis expresses the role of climate in dynamic equilibrium in terms highly reminiscent of the much better-known definition applied to grade in rivers by J. H. Mackin twenty-two years later.

Approximately one-third of this volume is given over to extended translations of the work of the outstanding modern exponents of the climatic-geomorphological approach: Julius Büdel in Germany and Jean Tricart and André Cailleux in France. This weighting can be justified on the grounds of their theoretical pre-eminence, the comprehensiveness of their approach and, in a collection of geographical readings, their concern with distributions and regions, as expressed in their presentation of data in map form.

Büdel's 1948 paper is an important attempt at regional synthesis on a world scale. It attempts to establish climatic-morphological zones by the regional grouping of distinctive landform assemblages, which in turn may be used to deduce the climates responsible for them; Büdel calls this the main aim of climatic geomorphology. The widespread modification of large areas of the earth during the Pleistocene is recognised and used as a datum in an attempt to assess to what extent the landform style of a region is the product of the prevailing climate. Aside from the difficulty that the validity of Büdel's main premise is yet to be established in any precise way, the result is extremely general and, as Tricart and Cailleux point out, the zonal criterion is not strictly applied.

The contribution by Tricart and Cailleux contrasts with Büdel's work in several important respects. For example, they recognise vegetation as a geomorphic factor which is sometimes of dominant status, as in the temperate zone. Their morphoclimatic regions of the world are based principally on variations in the process complex consisting of physical and biochemical reactions, so that the terminology of their classification is pedological and biogeographical rather than climatic and geomorphic. Regional boundaries, therefore, rarely coincide with single climatic or biogeographic parameters. Morphological zones are defined in terms of geomorphological assemblages whose character is derived from the processes typical of that zone. The emphasis is on present-day processes, but the importance of paleogeomorphic and paleoclimatic factors is recognised. Throughout the sections dealing with each morphological region in turn, the emphasis remains on bioclimate and geomorphic *process* rather than on the characterisation of the zonal landform assemblage in the manner of Büdel.

The generalisation made by Büdel in 1948 that landform contrasts due to climatic differences become evident only on a broad scale with the individual landforms diagnostic of a particular climate tending to be the smaller features of an area, and landforms due to aclimatic factors being, on the other hand, the larger features which dominate at more local scales, is further emphasised in his 1963 paper. Here, the effect of paleo-landforms is no longer treated as a qualification of a classification of current landform–climate relationships, but is considered essential to his climatogenetic subdivision of the earth into regions consisting of suites of dominant landforms. While the regional subdivision of the globe on such a basis is, as might be expected, highly generalised, Büdel's work remains of interest for at least two reasons: first, because of his attempt to use data drawn from his own wide field experience (especially in the periglacial and savanna regions) to delimit morphological processes on the ground; and second, for his recognition of the hierarchical relationship between process geomorphology, climatic geomorphology and climatogenetic geomorphology. The unsatisfactory nature of the latter, as represented by Büdel's 1963 paper, derives from the imperfect demonstration of the basic postulate of climatic geomorphology that climatic differences are expressed systematically in variations in particular subaerial landforms or landform suites. It is still true to say that, below the broad level of generalisation reached during the 1920s, the nature of this relationship remains essentially undefined. This situation arises from the fact that climatic geomorphology, in its turn, is dependent for its more precise definition on results derived from process geomorphology. While this fundamental field of study is showing vigorous growth, there remains as yet a paucity of information on the precise significance of climatic events to geomorphic processes and forms, the nature of the climatic component in variations in particular landforms, and the range of variation in landforms within an essentially uniform climatic environment. Traditional climatic geomorphology as represented by most of the papers in this volume has to a large extent glossed over this paucity of knowledge of fundamentals; it may be said to have proceeded, like Davis's work, to premature generalisation on the basis of quite vague ideas on the underlying process relations. In the light of this, the way forward in climatic geomorphology, already recognised in studies of fluvial catchments, becomes evident. It will involve the geomorphologist increasingly in sophisticated instrumentation of both climatic and geomorphological parameters, the design of specific, long-term experiments and the use of multivariate methods. The requisite technology, plant and techniques are available and await the formulation of appropriate experimental designs.

1 The Geographical Cycle

WILLIAM MORRIS DAVIS

THE GENETIC CLASSIFICATION OF LANDFORMS

ALL the varied forms of the lands are dependent on—or, as the mathematician would say, are functions of—three variable quantities, which may be called structure, process, and time. In the beginning, when the forces of deformation and uplift determine the structure and attitude of a region, the form of its surface is in sympathy with its internal arrangement, and its height depends on the amount of uplift that it has suffered. If its rocks were unchangeable under the attack of external processes, its surface would remain unaltered until the forces of deformation and uplift acted again; and in this case structure would be alone in control of form. But no rocks are unchangeable; even the most resistant yield under the attack of the atmosphere, and their waste creeps and washes downhill as long as any hills remain; hence all forms, however high and however resistant, must be laid low, and thus destructive process gains rank equal to that of structure in determining the shape of a land mass. Process cannot, however, complete its work instantly, and the amount of change from initial form is therefore a function of time. Time thus completes the trio of geographical controls, and is, of the three, the one of the most frequent application and of a most practical value in geographical description.

Structure is the foundation of all geographical classifications in which the trio of controls is recognised. The Allegheny plateau is a unit, a 'region', because all through its great extent it is composed of widespread horizontal rock layers. The Swiss Jura and the Pennsylvania Appalachians are units, for they consist of corrugated strata. The Laurentian highlands of Canada are essentially a unit, for they consist of greatly disturbed crystalline rocks. These geographical units have, however, no such simplicity as mathematical units; each one has a certain variety. The strata of plateaus are not strictly horizontal, for they slant or roll gently, now this way, now that. The corrugations of the Jura or of the Appalachians are not all alike; they might, indeed, be more truly described as all different, yet they preserve their essential features with much constancy. The disordered rocks of the Laurentian highlands have so excessively complicated a structure as at present to defy description, except item by item; yet, in spite of the free variations from a single structural pattern, it is legitimate and useful to look in a broad way at such a region, and to regard it as a structural unit. The forces by which structures and attitudes have been determined do not come within the scope of geographical inquiry, but the structures acquired by the action of these forces serve as the essential basis for the genetic classification of geographical forms. For the purpose of this article, it will suffice to recognise two great structural groups: first, the group of horizontal structures, including plains, plateaus, and their derivatives, for which no single name has been suggested; second, the group of disordered structures, including mountains and their derivatives, likewise without a single name. The second group may be more elaborately subdivided than the first.

The destructive processes are of great variety. There is the chemical action of air and water, and the mechanical action of wind, heat, and cold, of rain and snow, rivers and glaciers, waves and currents. But as most of the land surface of the earth is acted on chiefly by weather changes and running water, these will be treated as forming a normal group of destructive processes, while the wind of arid deserts and the ice of frigid deserts will be considered as climatic modifications of the norm, and set apart for particular discussion.

TIME AS AN ELEMENT IN GEOGRAPHICAL TERMINOLOGY

The amount of change caused by destructive processes increases with the passage of time, but neither the amount nor the rate of change is a simple function of time. The amount of change is limited, in the first place, by the altitude of a region above the sea; for, however long the time, the normal destructive forces cannot wear a land surface below this ultimate base-level of their action, and glacial and marine forces cannot wear down a land mass indefinitely beneath sea-level. The rate of change under normal processes, which alone will be considered for the present, is at the very first relatively moderate; it then advances rather rapidly to a maximum, and next slowly decreases to an indefinitely postponed minimum.

Evidently a longer period must be required for the complete denudation of a resistant than for that of a weak land mass, but no measure in terms of years or centuries can now be given to the period needed for the effective wearing down of highlands to feature-less lowlands. All historic time is hardly more than a negligible fraction of so vast a duration. The best that can be done at present is to give a convenient name to this unmeasured part of eternity, and for this purpose nothing seems more appropriate than a *geographical cycle*. When it is possible to establish a ratio between geographical and geological units, there will probably be found an approach to equality between the duration of an average cycle and that of Creta-ceous or Tertiary time, as has been indicated by the studies of several geomorphologists.

THE IDEAL GEOGRAPHICAL CYCLE

The sequence in the developmental changes of landforms is, in its own way, as systematic as the sequence of changes found in the more evident development of organic forms. Indeed, it is chiefly for this reason that the study of the origin of landforms—or geomorphogeny, as some call it—becomes a practical aid, helpful to the geographer at every turn. This will be made clearer by the specific consideration of an ideal case.

In Fig. 1.1 the base line αω represents the passage of time, while verticals above the base line measure altitude above sea-level. At the

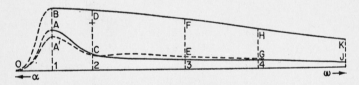

Fig. 1.1

epoch 1 let a region of whatever structure and form be uplifted, B representing the average altitude of its higher parts and A that of its lower parts, AB thus measuring its average initial relief. The surface rocks are attacked by the weather. Rain falls on the weathered surface and washes some of the loosened waste down the initial slopes to the trough lines, where two converging slopes meet; there the streams are formed, flowing in directions consequent upon the descent of the trough lines. The machinery of the destructive processes is thus put in motion, and the destructive development of the region is begun. The larger rivers, whose channels initially had an altitude

A, quickly deepen their valleys, and at the epoch 2 have reduced their main channels to a moderate altitude represented by C. The higher parts of the interstream uplands, acted on only by the weather without the concentration of water in streams, waste away much more slowly, and at epoch 2 are reduced in height only to D. The relief of the surface has thus been increased from AB to CD. The main rivers then deepen their channels very slowly for the rest of their lives, as shown by the curves CEGJ, and the wasting of the uplands, much dissected by branch streams, comes to be more rapid than the deepening of the main valleys, as shown by comparing the curves DFHK and CEGJ. The period 3–4 is the time of the most rapid consumption of the uplands, and thus stands in strong contrast with the period 1–2, when there was the most rapid deepening of the main valleys. In the earlier period the relief was rapidly increasing in value, as steep-sided valleys were cut beneath the initial troughs. Through the period 2–3 the maximum value of relief is reached, and the variety of form is greatly increased by the headward growth of side valleys. During the period 3–4 relief is decreasing faster than at any other time, and the slope of the valley sides is becoming much gentler than before; but these changes advance much more slowly than those of the first period. From epoch 4 onward the remaining relief is gradually reduced to smaller and smaller measures, and the slopes become fainter and fainter, so that some time after the last stage of the diagram the region is only a rolling lowland, whatever may have been its original height. So slowly do the later changes advance that the reduction of the reduced JK to half of its value might well require as much time as all that which has already elapsed; and from the gentle slopes that would then remain, the further removal of waste must indeed be exceedingly slow. The frequency of torrential floods and of landslides in young and in mature mountains, in contrast to the quiescence of the sluggish streams and the slow movement of the soil on lowlands of denudation, suffices to show that rate of denudation is a matter of strictly geographical as well as of geological interest.

It follows from this brief analysis that a geographical cycle may be subdivided into parts of unequal duration, each one of which will be characterised by the degree and variety of the relief, and by the rate of change, as well as by the amount of change that has been accomplished since the initiation of the cycle. There will be a brief youth of rapidly increasing relief, a maturity of strongest relief and greatest variety of form, a transition period of most rapidly yet slowly decreasing relief, and an indefinitely long old age of faint relief, in which further changes are exceedingly slow. There are, of course, no breaks between these subdivisions or stages; each one merges into its successor, yet each one is in the main distinctly characterised by features found at no other time.

THE DEVELOPMENT OF CONSEQUENT STREAMS

The preceding section gives only the barest outline of the systematic sequence of changes that run their course through a geographical cycle. The outline must be at once gone over, in order to fill in the more important details. In the first place, it should not be implied, as was done in Fig. 1.1, that the forces of uplift and deformation act so rapidly that no destructive changes occur during their operation. A more probable relation at the opening of a cycle of change places the beginning of uplift at O (Fig. 1.1) and its end at 1. The divergence of the curves OB and OA then implies that certain parts of the disturbed region were uplifted more than others, and that from a surface of no relief at sea-level at epoch O, an upland having AB relief would be produced at epoch 1. But even during uplift the streams that gather in the troughs as soon as they are defined do some work, and hence young valleys are already incised in the trough bottoms when epoch 1 is reached, as shown by the curve OA. The uplands also waste more or less during the period of disturbance, and hence no absolutely unchanged initial surface should be found, even for some time before epoch 1. Instead of looking for initial divides, separating initial slopes which descend to initial troughs, followed by initial streams, such as were implied in Fig. 1.1 at the epoch of instantaneous uplift, we must always expect to find some greater or lesser advance in the sequence of developmental changes, even in the youngest known landforms. 'Initial' is therefore a term adapted to ideal rather than to actual cases, in treating which the term 'sequential' and its derivatives will be found more appropriate. All the changes which directly follow the guidance of the ideal initial forms may be called consequent; thus a young form would possess consequent divides, separating consequent slopes, which descend to consequent valleys, the initial troughs being changed to consequent valleys in so far as their form is modified by the action of the consequent drainage.

THE GRADE OF VALLEY FLOORS

The larger rivers soon—in terms of the cycle—deepen their main valleys, so that their channels are but little above the base-level of the region; but the valley floor cannot be reduced to the absolute base-level, because the river must slope down to its mouth at the seashore. The altitude of any point on a well-matured valley floor must therefore depend on river slope and distance from mouth. Distance from mouth may here be treated as a constant, although a fuller statement would consider its increase in consequence of delta growth. As engineers know very well, river slope cannot be less than a certain minimum that is determined by both volume and quantity and texture of detritus or load. Volume may be temporarily taken as a constant, although it may easily be shown to suffer important changes during the progress of a normal cycle. Load is small at the beginning and rapidly increases in quantity and coarseness during youth, when the region is entrenched by steep-sided valleys; it continues to increase in quantity, but probably not in coarseness, during early maturity, when ramifying valleys are growing by headward erosion and are thus increasing the area of wasting slopes; but after full maturity, load continually decreases in quantity and in coarseness of texture, and during old age the small load that is carried must be of very fine texture or else must go off in solution. Let us now consider how the minimum slope of a main river will be determined.

In order to free the problem from unnecessary complications, let it be supposed that the young consequent rivers have at first slopes that are steep enough to make them all more than competent to carry the load that is washed into them from the wasting surface on either side, and hence competent to entrench themselves beneath the floor of the initial troughs. This is the condition tacitly postulated in Fig. 1.1.

If a young consequent river be followed from end to end, it may be imagined as deepening its valley everywhere, except at the very mouth. Valley deepening will go on most rapidly at some point, probably nearer head than mouth. Above this point the river will find its slope increased; below, decreased. Let the part upstream from the point of most rapid deepening be called the headwaters, and the part downstream the lower course or trunk. In consequence of the changes thus systematically brought about, the lower course of the river will find its slope and velocity decreasing and its load increasing, that is, its ability to do work is becoming less, while the work that it has to do is becoming greater. The original excess of ability over work will thus in time be corrected, and when an equality of these two quantities is brought about, the river is *graded*, this being a simple form of expression, suggested by Gilbert, to replace the more cumbersome phrases that are required by the use of 'profile of equilibrium' of French engineers. When the graded condition is reached, alteration of slope can take place only as volume and load change their relation; and changes of this kind are very slow.

In a land mass of homogeneous texture the graded condition of a river would be, in such cases as are above considered, first attained at the mouth, and would then advance retrogressively upstream. When the trunk streams are graded, early maturity is reached; when the smaller headwaters and side streams are also graded, maturity is far advanced; and when even the wet-weather rills are graded, old age is attained. In a land mass of heterogeneous texture the rivers will be divided into sections by the belts of weaker and stronger rocks that they traverse; each section of weaker rocks will in due time be graded with reference to the section of harder rock next downstream, and thus the river will come to consist of alternating quiet reaches and hurried falls or rapids. The less resistant of the harder rocks will be slowly worn down to grade with respect to the more resistant ones that are farther downstream; thus the rapids will decrease in number, and only those on the very strongest rocks will long survive. Even these must vanish in time, and the graded condition will then be extended from mouth to head. The slope that is adopted when grade is assumed varies inversely with the volume; hence rivers retain steep headwaters long after their lower course is worn down almost level; but in old age even the headwaters must have a gentle declivity and moderate velocity, free from all torrential features. The so-called 'normal river', with torrential headwaters and a well-graded middle and lower course, is therefore simply a maturely developed river. A young river may normally have falls, even in its lower course, and an old river must be free from rapid movement even near its head.

If an initial consequent stream is for any reason incompetent to carry away the load that is washed into it, it cannot degrade its channel, but must aggrade instead (to use an excellent term suggested by Salisbury). Such a river then lays down the coarser part of the offered load, thus forming a broadening flood land, building up its valley floor, and steepening its slope until it gains sufficient velocity to do the required work. In this case the graded condition is reached by filling up the initial trough instead of by cutting it down. Where basins occur, consequent lakes rise in them to the level of the outlet at the lowest point of the rim. As the outlet is cut down, it forms a sinking, local base-level, with respect to which the basin is aggraded; and as the lake is thus destroyed, it forms a sinking base-level, with respect to which the tributary streams grade their valleys; but, as in the case of falls and rapids, the local base-levels of outlet and lake are temporary and lose their control when the main drainage lines are graded with respect to absolute base-level in early or late maturity.

THE DEVELOPMENT OF RIVER BRANCHES

Several classes of side streams may be recognised. Some of them are defined by slight initial depressions in the side slopes of the main river troughs; these form lateral or secondary consequents, branching from a main consequent, generally running in the direction of the dip of the strata. Others are developed by headward erosion under the guidance of weak substructures that have been laid bare on the valley walls of the consequent streams; they follow the strike of the strata and are entirely regardless of the form of the initial land surface; they may be called subsequent, this term having been used by Jukes in describing the development of such streams. Still others develop here and there, to all appearances by accident, seemingly independent of systematic guidance; these are common in horizontal or massive structures. While waiting to learn just what their control may be, their independence of apparent control may be indicated by calling them insequent.

RELATION OF RIVER ABILITY AND LOAD

As the dissection of a land mass proceeds with the fuller development of its consequent, subsequent, and insequent streams, the area of steep valley sides greatly increases from youth into early and full maturity. The waste that is delivered by the side branches to the main stream comes chiefly from the valley sides, and hence its quantity increases with the increase of strong dissection, reaching a maximum when the formation of new branch streams ceases, or when the decrease in the slope of the wasting valley sides comes to balance their increase of area. It is interesting to note in this connection the consequences that follow from two contrasted relations of the date for the maximum discharge of waste and of that for the grading of the trunk streams. If the first is not later than the second, the graded rivers will slowly assume gentler slopes as their load lessens; but as the change in the discharge of waste is almost infinitesimal compared to the amount discharged at any one time, the rivers will essentially preserve their graded condition in spite of the minute excess of ability over work. On the other hand, if the maximum of load is not reached until after the first attainment of the graded condition by the trunk rivers, then the valley floors will be aggraded by the deposition of a part of the increasing load, and thus a steeper slope and a greater velocity will be gained whereby the remainder of the increase can be borne along. The bottom of the V-shaped valley, previously carved, is thus slowly filled with a gravelly flood plain, which continues to rise until the epoch of the maximum load is reached, after which the slow degradation indicated above is entered upon. Early maturity may therefore witness a slight shallowing of the main valleys instead of the slight deepening (indicated by the dotted line CE in Fig. 1.1); but late maturity and all old age will be normally occupied by the slow continuation of valley erosion that was so vigorously begun during youth.

THE DEVELOPMENT OF RIVER MEANDERS

It has been thus far implied that rivers cut their channels vertically downward, but this is far from being the whole truth. Every turn in the course of a young consequent stream causes the stronger currents to press toward the outer bank, and each irregular, or perhaps subangular, bend is thus rounded out to a comparatively smooth curve. The river therefore tends to depart from its irregular initial path towards a serpentine course, in which it swings to right and left over a broader belt than at first. As the river cuts downwards and outwards at the same time, the valley slopes become unsymmetrical being steeper on the side toward which the current is urged by centrifugal force. The steeper valley side thus gains the form of a half-amphitheatre, into which the gentler-sloping side enters as a spur of the opposite uplands.

When the graded condition is attained by the stream, downward cutting practically ceases, but outward cutting continues; a normal flood plain is then formed as the channel is withdrawn from the gently sloping side of the valley. Flood plains of this kind are easily distinguished in their early stages from those already mentioned (formed by aggrading the flat courses of incompetent young rivers, or by aggrading the graded valleys of overloaded rivers in early maturity); for these occur in detached lunate areas of the stream, first on one side, then on the other side, and always systematically placed at the foot of the gentler-sloping spurs. But as time passes, the river impinges on the upstream side and withdraws from the downstream side of every spur, and thus the spurs are gradually consumed. They are first sharpened, so as better to deserve their name; they are next reduced to short cusps; then they are worn back to blunt salients; and, finally, they are entirely consumed, and the river wanders freely on its open flood plain, occasionally swinging against the valley side. By this time the curves of youth are changed into systematic meanders, of radius appropriate to river volume; and for all the rest of an undisturbed life the river persists in the habit of serpentine flow. The less the slope of the flood plain becomes in advancing old age, the larger the arc of each meander, and hence the longer the course of the river from any point to its mouth. Increase of length from this cause must tend to diminish fall, and thus to render the river less competent than it was before. The result of this tendency will be to retard the already slow process by which a gently sloping flood plain is degraded, so as to approach coincidence with a level surface; but it is not likely that old rivers often remain undisturbed long enough for the full realisation of these theoretical conditions.

The migration of divides must now and then result in a sudden increase in the volume of one river and in a correspondingly sudden decrease of another. After such changes, accommodation to the changed volume must be made in the meanders of each river affected. The one that is increased will call for enlarged dimensions; it will usually adopt a gentler slope, thus terracing its flood plain, and demand a greater freedom of swinging, thus widening its valley. The one that is decreased will have to be satisfied with smaller dimensions; it will wander aimlessly in relatively minute meanders on its flood plain, and from increase of length, as well as from loss of volume, it will become incompetent to transport the load brought in by the side streams, and thus its flood plain must be aggraded. There are beautiful examples known of both of these peculiar conditions.

THE DEVELOPMENT OF GRADED VALLEY SIDES

When the migration of divides ceases in late maturity, and the valley floors of the adjusted streams are well graded, even far toward the headwaters, there is still to be completed another and perhaps even more remarkable sequence of systematic changes than any yet described: this is the development of graded waste slopes on the valley sides. It is briefly stated that valleys are eroded by their rivers, yet there is a vast amount of work performed in the erosion of valleys in which rivers have no part. It is true that rivers deepen the valleys in the youth and widen the valley floors during the maturity and old age of a cycle, and that they carry to the sea the waste denuded from the land; it is this work of transportation to the sea that is peculiarly the function of rivers, but the material to be transported is supplied chiefly by the action of the weather on the steeper conse-quent slopes and on the valley sides. The transportation of the weathered material from its source to the stream in the valley bottom is the work of various slow-acting processes, such as the surface wash of rain, the action of ground-water, changes of tempera-ture, freezing and thawing, chemical disintegration and hydration, the growth of plant roots, the activities of burrowing animals. All these cause the weathered rock waste to wash and creep slowly downhill, and in the motion thus ensuing there is much that is analogous to the flow of a river. Indeed, when considered in a very broad and general way, a river is seen to be a moving mixture of water and waste in variable proportions, but mostly water; while a creeping sheet of hillside waste is a moving mixture of waste and water in variable proportions, but mostly waste. Although the river and the hillside waste sheet do not resemble each other at first sight, they are only the extreme members of a continuous series, and when this generalisation is appreciated, one may fairly extend the 'river' all over its basin and up to its very divides. Ordinarily treated, the river is like the veins of a leaf; broadly viewed, it is like the entire leaf. The validity of this comparison may be more fully accepted when the analogy, indeed the homology, of waste sheets and water streams is set forth.

In the first place, a waste sheet moves fastest at the surface and slowest at the bottom, like a water stream. A graded waste sheet may be defined in the very terms applicable to a graded water stream; it is one in which the ability of the transporting forces to do work is equal to the work that they have to do. This is the condition that obtains on those evenly slanting, waste-covered mountain sides which have been reduced to a slope that engineers call 'the angle of repose', because of the apparently stationary condition of the creeping waste, but that should be called, from the physiographic standpoint, 'the angle of first-developed grade'. The rocky cliffs and ledges that often surmount graded slopes are not yet graded; waste is removed from them faster than it is supplied by local weathering and by creeping from still higher slopes, and hence the cliffs and ledges are left almost bare; they correspond to falls and rapids in water streams, where the current is so rapid that its cross-section is much reduced. A hollow on an initial slope will be filled to the angle of grade by waste from above; the waste will accumulate until it reaches the lowest point on the rim of the hollow, and then outflow of waste will balance inflow. Here is the evident homologue of a lake.

In the second place, it will be understood, from what has already been said, that rivers normally grade their valleys retrogressively from the mouth headwards, and that small side streams may not be graded till long after the trunk river is graded. So with waste sheets; they normally begin to establish a graded condition at their base and then extend it up the slope of the valley side whose waste they 'drain'. When rock masses of various resistance are exposed on the valley side, each one of the weaker is graded with reference to the stronger one next downhill, and the less resistant of the stronger ones are graded with reference to the more resistant, or with reference to the base of the valley side; this is perfectly comparable to the development of graded stretches and to the extinction of falls and rapids in rivers. Ledges remain ungraded on ridge crests and on the convex front of hill spurs long after the graded condition is reached in the channels of wet-weather streams in the ravines between the spurs; this corresponds nicely with the slower attainment of grade in small side streams than in large trunk rivers. But as late maturity passes into old age, even the ledges on ridge crests and spur fronts disappear, all being concealed in a universal sheet of slowly creeping waste. From any point on such a surface a graded slope leads the waste down to the streams. At any point the agencies of removal are just able to cope with the waste that is there weathered *plus* that which comes from farther uphill. This wonderful condition is reached in certain well-denuded mountains, now subdued from their mature vigour to the rounded profiles of incipient old age. When the full meaning of their graded form is apprehended, it constitutes one of the strongest possible arguments for the sculpture of the lands by the slow processes of weathering long continued. To look upon a landscape of this kind without any recognition of the labour expended in producing it, or of the extraordinary adjustments of streams to structures and of waste to weather, is like visiting Rome in the ignorant belief that the Romans of today have had no ancestors.

Just as graded rivers slowly degrade their courses after the period of maximum load is past, so graded waste sheets adopt more and more gentle slopes when the upper ledges are consumed and coarse waste is no longer plentifully shed to the valley sides below. A changing adjustment of a most delicate kind is here discovered. When the graded slopes are first developed they are steep, and the waste that covers them is coarse and of moderate thickness; here the strong agencies of removal have all they can do to dispose of the plentiful supply of coarse waste from the strong ledges above, and the no less plentiful supply of waste that is weathered from the weaker rocks beneath the thin cover of detritus. In a more advanced stage of the cycle the graded slopes are moderate, and the waste that covers them is of finer texture and greater depth than before; here the weakened agencies of removal are favoured by the slower weathering of the rocks beneath the thickened waste cover, and by the greater refinement (reduction to finer texture) of the loose waste during its slow journey. In old age, when all the slopes are very gentle, the agencies of waste removal must everywhere be weak, and their equality with the processes of waste supply can be maintained only by the reduction of the latter to very low values. The waste sheet then assumes a great thickness—even fifty or a hundred feet—so that the progress of weathering is almost nil; at the same time, the surface waste is reduced to an extremely fine texture, so that some of its particles may be moved even on faint slopes. Hence the occurrence of deep soils is an essential feature of old age, just as the occurrence of bare ledges is of youth. The relationships here obtaining are as significant as those which led Playfair to his famous statement concerning the origin of valleys by the rivers that drain them.

OLD AGE

Maturity is passed and old age is fully entered upon when the hilltops and the hillsides, as well as the valley floors, are graded. No new features are now developed, and those that have been earlier developed are weakened or even lost. The search for weak structures and the establishment of valleys along them has already been thoroughly carried out; now the larger streams meander freely in open valleys and begin to wander away from the adjustments of maturity. The active streams of the time of greatest relief now lose their headmost branches, for the rainfall is lessened by the destruction of the highlands, and the runoff of the rainwater is retarded by the flat slopes and deep soils. The landscape is slowly tamed from its earlier strength and presents only a succession of gently rolling swells alternating with shallow valleys, a surface everywhere open to occupation. As time passes, the relief becomes less and less; whatever the uplifts of youth, whatever the disorder and hardness of the rocks, an almost featureless plain (a peneplain), showing little sympathy with structure, and controlled only by a close approach of base-level, must characterise the penultimate stage of the uninterrupted cycle; and the ultimate stage would be a plain without relief.

Some observers have doubted whether even the penultimate stage of a cycle is ever reached, so frequently do movements in the earth's crust cause changes in its position with respect to base-level. But, on the other hand, there are certain regions of greatly disordered structure whose small relief and deep soils cannot be explained without supposing them to have, in effect, passed through all the stages above described (and doubtless many more, if the whole truth were told) before reaching the penultimate, whose features they verify. In spite of the great disturbances that such regions have suffered in past geological periods, they have afterwards stood still so long, so patiently, as to be worn down to peneplains over large areas, only here and there showing residual reliefs where the most resistant rocks still stand up above the general level. Thus verification is found for the penultimate, as well as for many earlier stages, of the ideal cycle.

INTERRUPTIONS OF THE IDEAL CYCLE

One of the first objections that might be raised against a terminology based on the sequence of changes through the ideal uninterrupted cycle is that such a terminology can have little practical application on an earth whose crust has the habit of rising and sinking frequently during the passage of geological time. To this it may be answered that if the scheme of the geographical cycle were so rigid as to be incapable of accommodating itself to the actual condition of the earth's crust, it would certainly have to be abandoned as a theoretical abstraction; but such is by no means the case. Having traced the normal sequence of events through an ideal cycle, our next duty is to consider the effects of any and all kinds of movements of the land mass with respect to its base-level. Such movements must be imagined as small or great, simple or complex, rare or frequent, gradual or rapid, early or late. Whatever their character, they will be called 'interruptions', because they determine a more or less complete break in processes previously in operation, by beginning a new series of processes with respect to the new base-level. Whenever interruptions occur, the pre-existent conditions that they interrupt can be understood only after one has analysed them in accordance with the principles of the cycle, and herein lies one of the most practical applications of what at first seems remotely theoretical. A land mass, uplifted to a greater altitude than it had before, is at once more intensely attacked by the denuding processes in the new cycle thus initiated; but the forms on which the new attack is made can only be understood by considering what had been accomplished in the preceding cycle previous to its interruption. It will be possible here to consider only one or two specific examples from among the multitude of interruptions that may be imagined.

Let it be supposed that a maturely dissected land mass is evenly uplifted five hundred feet above its former position. All the graded streams are hereby revived to new activities, and proceed to entrench their valley floors, in order to develop graded courses with respect to the new base-level. The larger streams first show the effect of the change; the smaller streams follow suit as rapidly as possible. Falls reappear for a time in the river channels and then are again worn away. Adjustments of streams to structures are carried farther in the second effort of the new cycle than was possible in the single effort of the previous cycle. Graded hillsides are undercut; the waste washes and creeps down from them, leaving a long, even slope of bare rock; the rocky slope is hacked into an uneven face by the weather until at last a new graded slope is developed. Cliffs that had been extinguished on graded hillsides in the previous cycle are thus for a time brought to life again, like the falls in the rivers, only to disappear in the late maturity of the new cycle.

The combination of topographical features belonging to two cycles may be called composite topography, and many examples could be cited in illustration of this interesting association. In every case, description is made concise and effective by employing a terminology derived from the scheme of the cycle. For example, Normandy is an uplifted peneplain, hardly yet in the mature stage of its new cycle; thus stated, explanation is concisely given to the meandering course of the rather narrow valley of the Seine, for this river has carried forward into the early stages of the new cycle the habit of swinging in strong meanders that it had learned in the later stages of the former cycle.

If the uplift of a dissected region be accompanied by a gentle tilting, then all the water streams and waste streams whose slope is increased will be revived to new activity, while all those whose slope is decreased will become less active. The divides will migrate into the basins of the less active streams, and the revived streams will gain length and drainage area. If the uplift be in the form of an arch, some of the weaker streams whose course is across the axis of the arch may be, as it were, 'broken in half'; a reversed direction of flow may thus be given to one part of the broken stream; but the stronger rivers may still persevere across the rising arch in spite of its uplift, cutting down their channels fast enough to maintain their direction of flow unchanged. Such rivers are known as antecedent rivers.

The changes introduced by an interruption involving depression are easily deduced. Among their most interesting features is the invasion of the lower valley floors by the sea, thus 'drowning' the valleys to a certain depth and converting them into bays. Movements that tend to produce trough-like depressions across the course of a river usually give birth to a lake of water or waste in the depressed part of the river valley. In mountain ranges frequent and various interruptions occur during the long period of deformation; the Alps show so many recent interruptions that a student there would find little use for the ideal cycle; but in mountain regions of ancient deformation the disturbing forces seem to have become almost extinct, and there the ideal cycle is almost realised. Central France gives good illustration of this principle. It is manifest that one might imagine an endless number of possible combinations among the several factors of structure, stage of development at time of interruption, character of interruption, and time since interruption.

ACCIDENTAL DEPARTURES FROM THE IDEAL CYCLE

Besides the interruptions that involve movements of a land mass with respect to base-level, there are two other classes of departure from the normal or ideal cycle that do not necessarily involve any such movements: these are changes of climate and volcanic eruptions, both of which occur so arbitrarily as to place and time that they may be called accidents. Changes of climate may vary from the normal toward the frigid or the arid, each change causing significant departures from normal geographical development. If a reverse change of climate brings back more normal conditions, the effects of the abnormal accident may last for some small part of a cycle's duration before they are obliterated. It is here that features of glacial origin belong, so common in north-western Europe and north-eastern America. Judging by the present analysis of glacial and interglacial epochs during Quaternary time, or of humid and arid epochs in the Great Salt Lake region, it must be concluded that accidental changes may occur over and over again within a single cycle.

The general scheme of the geographical cycle needs adaptation to two special climates: one is glacial, the other, arid.

THE CYCLE OF GLACIAL DENUDATION

The points of resemblance between rivers and glaciers, streams of water and streams of ice, are so numerous that they may be reasonably extended all through a cycle of denudation. Let us then inquire if glaciers may not, during their ideal life history, develop as orderly a succession of features as that which so well characterises the normal development of rivers.

Young glaciers will be those which have been just established in courses that are consequent upon the slopes of a newly uplifted land surface. Mature glaciers will be those which have eroded their valleys to grade and thus dissected the uplifted surface. Old glaciers will be those which cloak the whole lowland to which the upland has been reduced, or which are slowly fading in the milder climate of the low levels appropriate to the close of the cycle of denudation.

Imagine an initial land surface raised to a height of several hundred metres, with a moderate variety of relief due to deformation. Let the snowline stand at a height of 60 metres. As elevation progresses, snow accumulates on all the upland and highland surfaces. Glaciers are developed in every basin and trough; they creep slowly forward to lower ground, where they enter a milder climate (or the sea) and gradually melt away. At some point between its upper heads and its lower end, each glacier will have a maximum volume. Downstream from this point the glacier will diminish in size, partly by evaporation but more by melting, and the ice water thus provided will flow away from the end of the glacier in the form of an ordinary stream, carving its valley in normal fashion. Some erosion may be accomplished under the upper fields of snow and névé, but it is believed that more destructive work is done beneath the ice. The erosion is accomplished by weathering, scouring, plucking, and corrading. Weathering occurs where variations of external temperature penetrate to the bedrock, as is particularly the case between the séracs of glacial cascades, and again along the line of deep crevasses, or *bergschrunds*, that are usually formed around the base of reservoir walls, which are thus transformed into corries (cirques, *karen, botner*), as has been suggested by several observers. Scouring is the work of rock waste dragged along beneath the glacier, by which the bedrock is ground down, striated, and smoothed. Plucking results from friction under long-lasting heavy pressure, by which blocks of rock are removed bodily from the glacier bed and banks. Corrading is the work of subglacial streams, which must be well-charged with tools, large and small, and which must often flow under heavy pressure and with great energy. All these processes are here taken together as *glacial erosion*.

Let it be assumed that at first the slope of a glacier's path was steep enough to cause it to erode for the greater part or for the whole of its length. It is to be understood that a valley includes the channel that is eroded along its floor. The channel, with its beds and banks, is therefore that part of a valley which is occupied by the stream. Each young glacier will then proceed to cut down its consequent valley at a rate dependent on various factors, such as depth and velocity of ice stream, character of rock bed, quantity of ice-dragged waste, and so on; and the eroded channel in the bottom of the valley will in time be given a depth and width that will better suit the needs of ice discharge than did the initial basin or trough of the uplifted surface. The upper slopes of the glacial stream will thus be steepened, while its lower course will be given a gentler descent. Owing to the diminution of the glacier toward its lower end, the channel occupied by it will diminish in depth and breadth downwards from the point of maximum volume, this being analogous to the decrease in the size of the channel of a withering river below the point of its maximum volume. A time will come when all the energy of the glacier on its gentler slope will be fully taxed in moving forward the waste that has been brought down from the steeper slopes; then the glacier becomes only a transporting and not an eroding agent in its lower course. This condition will be first reached near the lower end, and slowly propagated headwards. Every part of the glacier in which the balance between ability to do work and work to be done is thus struck may be said to be 'graded'; and in all such parts the surface of the glacier will have a smoothly descending slope. Maturity will be reached when, as in the analogous case of a river, the nice adjustment between ability and work is extended to all parts of a glacial system. In the process of developing this adjustment a large trunk glacier might entrench the main valley more rapidly than one of the smaller branches could entrench its side valley; then for a time the branch would join the trunk in an ice rapid of many séracs. But when the trunk glacier had deepened its valley so far that further deepening became slow, the branch glacier would have opportunity to erode its side valley to an appropriate depth, and thus to develop an accordant junction of trunk and branch ice *surfaces*, although the *channels* of the larger and the smaller streams might still be of very unequal depth, and the channel *beds* might stand at discordant levels. If the glaciers should disappear at this stage of the cycle, their channels would be called valleys, and the discordance of the channel beds might naturally excite surprise. The few observers who, previous to 1898, commented upon a discordance of this kind explained it as a result of excessive erosion of the main valley by the trunk glacier, while the hanging lateral valleys were implicitly, if not explicitly, regarded as hardly changed from their preglacial form.

When the trunk and branch glaciers have developed well-defined, maturely graded valleys, the continuous snow mantle that covered the initial uplands of early youth is exchanged for a discontinuous cover, rent on the steep valley sides where weathering comes to have a greatly increased value, and thickened where the ice streams have

established their courses. This change corresponds to that between
the ill-defined initial drainage in the early youth of the river cycle and
the well-defined drainage in its maturity.

BROKEN-BEDDED VALLEYS

It is probable that variations in rock structure will have permitted a
more rapid development of the graded condition in one part of the
glacial valley than in another, as is the case with rivers of water.
Steady flowing reaches and broken rapids will thus be produced in
the ice stream during its youth; and the glacial channel may then be
described as *broken-bedded*. But all the rapids must be worn down
and all the reaches must become confluent in maturity. It is eminently
possible that the reaches on the weaker or more jointed rocks may be
eroded during youth to a somewhat greater depth than the sill of
more resistant or less jointed rock which is next downstream; and if
the glacier should vanish by climatic change while in this condition, a
lake would occupy the deepened reach, while the lake outlet would
flow forward over rocky ledges to the next lower reach or lake.
Many Norwegian valleys today seem to be in this condition. Indeed,
some observers have described broken-bedded valleys as the normal
product of glacial erosion, without reference to the early stage in the
glacial cycle of which broken-bedded glacial channels seem to be
characteristic. Truly, it is not always explicitly stated that the resistance of the rock bed varies appropriately to the change of form in a
broken-bedded channel, but the variations of structural resistance or
firmness that the searching pressure and friction of a heavy glacier
could detect might be hardly recognisable to our superficial observations; and, on the other hand, the analogy of young ungraded glaciers
with young ungraded rivers seems so natural and reasonable that
broken-bedded glacial channels ought to be regarded only as features
of young glacial action, not as persistent features always to be associated with glacial erosion. If the glaciers had endured longer in
channels of this kind, the 'rapids' and other inequalities by which the
bed may be interrupted must have been worn back and lowered, and
in time destroyed.

VALLEY SILLS

If a young glacier erodes its valley across rocks of distinctly different
resistances, a strong inequality of channel bed may be developed.

Basins of a considerable depth may be excavated in the weaker strata, while the harder rocks are less eroded and cross the valleys in rugged sills. Forms of this kind are known in Alpine valleys, for example, in the valley of the Aar above Meiringen and in the lower Gasternthal near its junction with the Kanderthal; in both these cases the basins have been aggraded and the sills have been trenched by the postglacial streams. In the lower Gasternthal the height and steepness of the rocky sill, when approached from upstream, is astonishing; its contrast to the basin that it encloses is difficult enough to explain even for those who are willing to accept strong glacial erosion. It should, however, be noted that river channels also are deeper in the weaker rocks upstream from a hard rock sill; if the river volume should greatly decrease, a small lake would remain above the sill, drained by a slender stream cutting a gorge through the sill.

If an initial depression occurred on the path of the glacier, so deep that the motion of the ice through it was much retarded, an ice lake would gather in it. Then the waste dragged into the basin from upstream might accumulate upon its floor until the depth of the basin was sufficiently decreased and the velocity of the ice through it sufficiently increased to bring about a balance between ability to do work and work to be done. Here the maturely graded condition of the ice stream would have been attained by aggrading its bed instead of degrading it; this being again closely analogous to the case of a river which aggrades initial depressions and degrades initial elevations in producing its maturely graded course.

ICE STREAMS

Water streams subdivide toward the headwaters into a great number of very fine rills, each of which may retrogressively cut its own ravine in a steep surface, not cloaked by waste. But the branches of a glacial drainage system are much more clumsy, and the channels that they cut back into the upland or mountain mass are round-headed or amphitheatre-like; but the beds of the branching glaciers cannot be cut as deep as the bed of the large glacial channel into which they flow; thus corries, perched on the side walls of large valleys, may be produced in increasing number and strength as glacial maturity approaches, and in decreasing strength and number as maturity passes into old age. As maturity approaches, the glacial system will include not only those branches that are consequent upon the initial form, but certain others which have come into existence by the headward erosion of their névé reservoirs following the guidance of weak structures; thus a maturely developed glacial drainage system may have its subsequent as well as its consequent branches. It is entirely conceivable, as has been suggested by Meunier, that one ice stream may capture the upper part of another. The conditions most favourable for such a process resemble those under which river diversions and adjustments take place, namely, a considerable initial altitude of the region, allowing a deep dissection; a significant difference of drainage areas or of slopes, whereby certain glaciers incise deeper valleys than others; a considerable diversity of mountain structure, permitting such growth and arrangement of subsequent glaciers as shall bring the head reservoir of a subsequent ice stream alongside of and somewhat beneath the banks of a consequent ice stream. Thus glacial systems may come to adjust their streams to the structures upon which they work, just as happens in river systems.

The load transported by a glacial system may at first be supplied largely by waste plucked and scoured from the beds of the glacial channels as well as by waste detached from the enclosing slopes; but in time, when the graded condition of the chief channels is reached and their further deepening almost ceases, by far the largest share of load will be supplied from the subaerial valley sides, where weathering of the ordinary kind will ravine the slopes, thus producing a topography that is strongly contrasted with the smooth walls of the glacial channels. If the initial glacial system should incise its channels so deeply beneath a lofty highland that the supply of waste from the valley sides continued to increase after the development of graded glacial channels, it is conceivable that the channel beds might have to be aggraded for a time, as is believed to be the case with river channels under similar conditions; but owing to the receipt on the glacial surface of waste from the valley sides, it is also conceivable that this analogy may not closely obtain. Toward the end of the ice stream it may well happen that the diminution of its volume and the consequent diminution of its capacity to do work will result in the aggradation of its bed by waste that cannot be carried farther forward. At the same time, the outflowing river may be unable to wash away all the waste that is delivered to it, and so, for a time through later youth and early maturity, the river may act as an aggrading agent and build up a broad, flat alluvial fan, such as fronts the terminal moraines of the Alpine glaciers that once descended to the plain of Lombardy. Some response to the change thus produced in the altitude of the end of the glacier may be expected far up its channel, whose bed would thus come to be aggraded with till. Similarly, the ice sheets that spread from the Scandinavian and Laurentian highlands over the lowlands in the south changed their behaviour from degrading agents in the central area to aggrading agents in the peripheral area. Hence, a belief in effective erosion is not antagonistic to a belief in effective deposition in the case of glaciers any more than in the case of rivers. In each case the action varies appropriately to its place in the drainage system and to its stage in the cycle. But there will be a later stage, when the wasting of the superglacial slopes reduces them to moderate declivity, so that the waste delivered from them decreases in quantity. Then the outflowing water stream at the end of the glacier may become a degrading agent, the altitude of the end of the glacier may be slowly lessened, and a very slow and long-continued deepening of the whole glacial channel will take place, without requiring a departure from an essentially graded condition.

As the general denudation of the region progresses, the snowfall must be decreased and the glacial system must shrink somewhat, leaving a greater area of lowland surface to ordinary river drainage. When the upland surface is so far destroyed that even the hilltops stand below the 60-metre contour, the snowfields will be represented only by the winter snowsheet, and the glaciers will have disappeared, leaving normal agencies to complete the work of denudation that they have so well begun.

THE ARID CLIMATE

There are certain essential features of the arid climate. Primarily there must be so little rainfall that plant growth is scanty, that no basins of initial deformation are filled to overflowing, that no large trunk rivers are formed, and hence that the drainage does not reach the sea.

The agencies of sculpture and their opportunities for work in arid regions are peculiar in several respects. The small rainfall and the dry air reduce the groundwater to a minimum. In its absence, weathering is almost limited to the surface, and is more largely physical than chemical. The streams are usually shorter than the slopes, and act as discontinuously at their lower as at their upper ends. The scarcity of plant growth leaves the surface relatively free to the attack of the winds and the intermittent waters. Hence, in the production of fine waste, the splitting, flaking, and splintering of local weathering are supplemented rather by the rasping and trituration that go with transportation than by the chemical disintegration that characterises a plant-bound soil.

No special conditions need be postulated as to the initiation of the arid cycle. The passive earth's crust may be (relatively) uplifted and offered to the sculpturing agencies with any structure, any form, and any altitude, in dry as well as in moist regions.

INITIAL STAGE

Let consideration be given to an uplifted region of large extent over which an arid climate prevails. Antecedent rivers, persisting from a previous cycle against the deformations by which the new cycle is introduced, must be rare, because such rivers should be large, and large rivers are unusual in an arid region. Consequent drainage must prevail. The initial slopes in each basin will lead the wash of local rains toward the central depression, whose lowest point serves as the local base-level for the district. There will be as many independent centripetal systems as there are basins of initial deformation, for no basin can contain an overflowing lake whose outlet would connect two centripetal systems. The centripetal streams will not always follow the whole length of the centripetal slopes; most of the streams of each basin system will wither away after descending from the less arid highlands to the more arid depressions. Each basin system will therefore consist of many separate streams, which may occasionally, in time of flood or in the cooler season of diminished evaporation, unite in an intermittent trunk river, and even form a shallow lake in the basin bed, but which will ordinarily exist independently as disconnected headwater branches.

YOUTHFUL STAGE

In the early stage of a normal cycle the relief is ordinarily and rapidly increased by the incision of consequent valleys by the trunk rivers that flow to the sea. In the early stage of the arid cycle the relief is slowly diminished by the removal of waste from the highlands, and its deposition on the lower gentler slopes and on the basin beds of all the separate centripetal drainage systems. Thus all the local base-levels rise. The areas of removal are in time dissected by valleys of normal origin; if the climate is very arid, the uplands and slopes of these areas are either swept bare, or left thinly veneered with angular stony waste, from which the finer particles are carried away almost as soon as they are weathered; if a less arid climate prevails on the uplands and highlands, the plants that they support will cause the retention of a larger proportion of finer waste on the slopes. The areas of deposition are, on the other hand, given a nearly level central floor of fine waste, with the varied phenomena of shallow lakes, *playas*, and *salinas*, surrounded with graded slopes of coarser waste. The deposits thus accumulated will be of variable composition and, toward the margin, of irregular structure. The coarser deposits will exhibit a variety of materials, mechanically comminuted, but not chemically disintegrated, and hence in this respect unlike the less heterogeneous deposits of humid climates from which the more easily soluble or decomposable minerals have been largely removed. The finer deposits will vary from sand and clay to salt and gypsum. The even strata that are supposed to characterise lake deposits may follow or precede irregular or cross-bedded strata, as the lake invades or is invaded by the deposits of streams or winds. While many desert deposits may be altogether devoid of organic remains, others may contain the fossils of land, stream, or lake organisms.

WIND ACTION

Streams, floods, and lakes are the chief agencies in giving form to the aggraded basin floors, as well as to the dissected basin margins in the early stages of the cycle. But the winds also are of importance; they do a certain share of erosion by sand-blast action. They do a more important work of transportation by sweeping the granular waste from exposed uplands and depositing it in more sheltered depressions, and by raising the finer dust high in the air and carrying it far and wide before it is allowed to settle. Wind action is, moreover, peculiar in not being guided by the slopes or restrained by the divides which control streams and stream systems. It is true that the winds, like the streams, tend in a very general way to wear down the highlands and to fill up the basins; but sand may be drifted uphill—dunes may be seen climbing strong slopes and escarpments in Arizona and Oregon—while fine dust carried aloft in whirlwinds and

dust storms is spread about by the upper currents with little regard
to the slopes of the land surface far below. Sand may be drifted, and
dust may be in this way carried outside the arid region from which
it was derived. Wind erosion may, furthermore, tend to produce
shallow depressions or hollows; for the whole region is the bed of the
wind, and is therefore to a certain extent analogous to the bed of a
river, where hollows are common enough; but in the early stages of
the cycle in a region where the initial relief was strong, the action
of the wind is not able to make hollows on the original slopes that
are actively worked upon, and for a time even steepened, by streams
and floods. Hence in the youthful stage windblown hollows are not
likely to be formed.

It is important to notice that a significant, though small, share
of windswept or windborne waste may be carried entirely outside of
or 'exported' from an arid region. It may be deposited on neighbouring lands, where it will be held among the grass of a less arid
climate, as long ago suggested by Richthofen; it may even be held
down on coastal lands by the dew, as has been suggested for certain
districts in Morocco by Fischer; it may fall into the sea, as is proved
by the sand that gives a ruddy tinge to the sails of vessels in the
Atlantic to leeward of the Sahara, and by the sand grains that are
dredged up with true pelagic deposits from the bottom of that part
of the Atlantic. It may therefore be expected that the progress of
erosion and waste exportation in a desert region will be associated
with the deposit of fine waste, as in loess sheets, on the neighbouring
less arid regions, especially down the course of the prevailing winds.
In regions of weak and variable winds the process of sand and dust
exportation must be extremely slow; in regions of steady winds it
must still be vastly slower than the ordinary rate of waste removal in
young or mature regions of plentiful rainfall and normal rivers. Yet
it is by this slow process of exportation that the mean altitude of an
arid region, such as is here considered, will be continually decreased;
hence the earlier stages of the arid cycle are predictably longer than
the corresponding stages of the normal cycle.

In the normal cycle the youthful stage is characterised by the
headward growth of many subsequent streams, chiefly along belts
of weak structures that are laid bare on the valley sides of the larger
consequent streams. In the arid cycle subsequent streams have a
smaller opportunity for development; first, because all the belts of
weak structure under the basin deposits are buried out of reach;

second, because in the absence of deep-cutting trunk rivers, many belts of weak structure are but little exposed. In so far, however, as the highlands are dissected by their headwater consequent streams, subsequent branches may grow out and diversify the slopes and rearrange the drainage.

MATURE STAGE

Continued erosion of the highlands and divides, and continued deposition in the basins, may here and there produce a slope from a higher basin floor across a reduced part of its initial rim to a lower basin floor. Headward erosion by the consequent or subsequent streams of the lower basin will favour this change, which might then be described as a capture of the higher drainage area. Aggradation of the higher basin is equally important, and a change thus effected might be described as an invasion of the lower basin by waste from the higher one; this corresponds in a belated way to the overflow of a lake in a normal cycle. There may still be no persistent stream connecting the two basins, but whenever rain falls on the slope that crosses the original divide, the wash will carry waste from the higher to the lower basin. Thus the drainage systems of two adjacent basins coalesce, and with this a beginning is made of the confluence and integration of drainage lines which, when more fully developed, characterise maturity. The intermittent drainage that is established across the former divide may have for a time a rather strong fall; as this is graded down to an even slope, an impulse of revival and deeper erosion makes its way, wave-like, across the floor of the higher basin and up all its centripetal slopes. The previously aggraded floor will thus for a time be dissected with a badland expression and then smoothed at a lower level, and the bordering waste slopes will be trenched and degraded, at the same time the lower basin floor will be more actively aggraded. If there is a sufficient difference of altitude between the two basins, all the waste that had been, in a preliminary or youthful view of the case, gathered in the higher basin, will in time be transferred to the lower basin; and thus a larger relation of drainage lines, a longer distance of intermittent transportation, a more continuous area of bedrock in the higher areas, and a more general concentration of waste in the lowest basins will be established. The higher local base-levels are thus, by a process of slow, inorganic natural selection, replaced by a smaller and smaller number of lower and lower base-levels; and with all this goes a headward extension of graded piedmont slopes, a deeper dissection of the highlands, and a better development of their subsequent and adjusted drainage. The processes of drainage adjustment are, however, at the best, of less importance here than in the normal cycle, because of the absence of main valleys, deep-cut by trunk rivers, and the resulting deficient development of deep-set subsequent streams, as has already been suggested.

As the coalescence of basins and the integration of stream systems progress, the changes of local base-levels will be fewer and slower and the obliteration of the uplands, the development of graded piedmont slopes, and the aggradation of the chief basins will be more and more extensive. The higher parts of the piedmont slopes may be rock floors, thinly and irregularly veneered with waste, as has been described by Keyes for certain basins (*bolsons*) in New Mexico; here, as well as upon the aggraded slopes and plains, sheet-flood action will prevail, as explained by McGee. The area occupied during early maturity by the three different kinds of surface—dissected highlands or mountains, graded piedmont slopes of rock or waste, and aggraded central plains with *playas*, *salinas*, or lakes—will depend on the initial relief, on the rock structure and its relation to desert weathering, on the percentage of material exported by the winds, and on the climate itself.

MATURITY AND WINDS

It is worth noting that, although the activity of streams and floods decreases with the decrease of relief and of slope, the activity of the winds is hardly affected as maturity advances. The winds do not depend on the gradient of the land surface for their gravitative acceleration; they may blow violently and work efficiently on a level surface. Whirlwinds are, indeed, most active on true plains. It may be that smooth plains are never swept by winds as violent as the blasts which attack highlands and mountains; but it is probable that the effective action of the winds is greater on a generally plain surface than on one of strong relief, where the salient ridges and peaks consist largely of firm rock, and where the loose waste is sheltered in re-entrant valleys. Moreover, it is in very great part on the plains that the winds of ordinary strength drift the sand about, and from the plains that whirlwinds and dust storms raise the finest waste high enough for exportation. It may therefore be concluded that the work of the winds is but little, if any, impaired by the general decrease of relief that characterises advancing maturity, and hence that their relative importance increases. Moreover, the scanty rainfall of an arid region will be decreased as its initial highlands, which originally acted as rain provokers, are worn down; hence, as the relief weakens, the winds will more and more gain the upper hand in the work of transportation. It is conceivable that the rate of exportation of sand and dust by the winds in maturity and all the later stages of an arid cycle is more rapid than the removal of fine soil, partly or largely in solution, from a plant-covered peneplain in the later stages of a normal cycle. Thus the slower work of the earlier stages of an arid cycle may be partly made good by the relatively more active work in the later stages.

As the processes thus far described continue through geological periods, the initial relief will be extinguished even under the slow processes of desert erosion, and there will appear instead large rock-floored plains sloping toward large waste-floored plains; the plains will be interrupted only where parts of the initial highlands and masses of unusually resistant rocks here and there survive as isolated residual mountains. At the same time, deposits of loess may be expected to accumulate in increasing thickness on the neighbouring less arid regions. The altitude at which the desert plain will stand is evidently independent of the general base-level, or sea-level, and dependent only on the original form and altitude of the region, and on the amount of dust that it has lost through wind transportation.

The most perfect maturity will be reached when the drainage of all the arid region becomes integrated with respect to a single aggraded basin base-level, so that the slopes lead from all parts of the surface to a single area for the deposition of the waste. The lowest basin area which thus comes to have a monopoly of deposition may receive so heavy a body of waste that some of its ridges may be nearly or quite buried. Strong relief may still remain in certain peripheral districts, but large plain areas will by this time necessarily have been developed. In so far as the plains are rock-floored, they will truncate the rocks without regard to their structure.

THE BEGINNING OF OLD AGE

During the advance of drainage integration the exportation of wind-borne waste is continued. At the same time the tendency of wind action to form hollows wherever the rocks weather most rapidly to a dusty texture would be favoured by the general decrease of surface slopes, and by the decrease of rainfall and of stream action resulting from the general wearing-down of the highlands. Thus it may well happen that windblown hollows are produced here and there, through the mature and later stages of the cycle, and that they will, even during early maturity, interfere to a greater or less degree with the development of the integrated drainage described above. In any case, it may be expected that windblown hollows will in late maturity seriously interfere with the maintenance of an integrated drainage system. Thus it appears that, along with the processes which tend toward the mature integration of drainage, there are other processes which tend toward a later disintegration, and that the latter gain efficiency as the former begin to weaken. A strong initial relief of large pattern, a quality of rock not readily reducible to dusty waste, and an irregular movement of light winds might give the control of sculpture to the intermittent streams through youth and into maturity; in such a case maturity might be characterised by a fully integrated system of drainage slopes, with insignificant imperfections in the way of windblown hollows. In a second region an initial form of weaker relief, a quality of rock readily reducible to dust, and a steady flow of strong winds might favour the development of windblown hollows or basins, and here the process of drainage disintegration would set in relatively early and prevent the attainment of mature drainage integration. In any case, as soon as the process of drainage disintegration begins to predominate, maturity may be said to pass into old age.

This feature of the arid cycle has no close analogy with the features recognised in the normal cycle. In the latter case the drainage systems of maturity tend on the whole to persist, even though the streams weaken and wander somewhat—and according to theory lose some of their adjustments—in very advanced old age. In the former case, as old age advances, the integrated and enlarged drainage systems of maturity are broken up into all manner of new, local, small and variable systems. The further results of drainage disintegration in the later stages of the cycle are even more peculiar.

PASSARGE'S GENERALISATION

As the drainage becomes more and more disintegrated, and the surface of the plain is slowly lowered, rock masses that most effectually resist dry weathering will remain as monadnocks—*Inselberge*, as Bornhardt and Passarge call them in South Africa. At the same time the waste will be washed away from the gathering grounds of maturity and scattered in the shallow hollows that are formed here and there by the winds as old age approaches. The removal of the basin deposits by the winds may be delayed where the hygroscopic action of saline clays keeps the surface firm; but wherever the integrated centripetal slopes are locally reversed by the hollowing action of the wind, some of the central deposits will be washed back again and exposed to renewed search for fine material by the wind, and thus a larger and larger part of the central waste will be redistributed and exported. As there is no relation of parts in the winds analogous to that of small branch and large trunk streams in river systems, the surface eroded by the winds need not slope toward any central area, but may everywhere be worn down essentially to the same level. The surface constantly wearing down, the waste washed irregularly about by the variable disintegration of the drainage system and continually exported by the winds, a nearly level rock floor, nowhere heavily covered with waste and everywhere slowly lowering at the rate of sand and dust exportation, is developed over a larger and larger area—such is the condition of quasi-equilibrium for old age. At last, as the waste is more completely exported, the desert plain may be reduced to a lower level than that of the deepest initial basin; and then a rock floor, thinly veneered with waste, unrelated to normal base-level, will prevail throughout, except where monadnocks still survive. This is the generalisation that we owe to Passarge; it seems to me secondary in value only to Powell's generalisation concerning the general base-level of erosion. Little wonder that an understanding of the possible development of rock-floored deserts of this kind, independent of base-level, was not reached inductively in western America; for there has been so much disturbance in the way of fracture and uplift in that region during Mesozoic, Tertiary, and Quaternary time that the attainment of arid old age has not been permitted.

If Passarge's views be now accepted, it follows that no truncated uplands should, without further inquiry, be treated as having been eroded when their region had a lower stand with respect to base-level; the possibility of their having been formed during an earlier arid climate as desert plains, without regard to the general base-level of the ocean, must be considered and excluded before base-levelling and uplift can be taken as proved.

It may at first appear sufficient to say that high-standing desert plains can have been made only in those regions which are now desert, but this easy solution of the problem is hardly convincing. Climatic changes are known to have occurred in the past, and inasmuch as they did not all affect areas in a way that is sympathetic with the present arrangement of the zones, the possibility of a former different distribution of deserts from that which now occurs seems to be open. Pleistocene climatic changes of the glacial kind were so modern and short-lived that they have little bearing on the possibility of earlier climatic changes of another order. The more ancient records of glaciation are so distributed as to demand significant rearrangement of the present climatic conditions. The existing deserts are, moreover, of two kinds with respect to cause: some deserts, like those of Africa and Australia, are arranged chiefly with respect to the trade-wind belt; other deserts, like those of central Asia and the south-western United States, are dependent for the most part on the extent and configuration of the surrounding highlands. When we go back as far as Cretaceous time, it should only be by evidence and not by assumption that we are led to regard a truncated upland of that date as having been base-levelled during a cycle of normal climate and afterward uplifted and dissected, instead of having been levelled above base-level during a cycle of arid climate, and dissected in consequence of a change to a normal climate.

Even if the climatic zones have always belted the earth as they do now, the desert areas that depend on the configuration of land and water, and of highlands and lowlands, have certainly varied through the geological ages. It is therefore desirable, wherever the question of 'uplifted and dissected peneplains' is raised, to scrutinise it carefully, and to determine, if possible, whether it is really the attitude of the earth's crust or the condition of climate that has been changed. It is likewise important to scrutinise desert plains, now standing above base-level, to see if they may not have been formed normally as lowland plains of erosion and afterward uplifted. It is therefore necessary to inquire into these features by which base-levelled peneplains and rock-floored desert plains may be distinguished, even though the former may be uplifted with a change to an arid climate, or though the latter may be depressed with a change to a humid climate.

2 Attempt at a Classification of Climate on a Physiographic Basis

ALBRECHT PENCK

IN place of the ancient's division of the earth's surface into parallel climatic zones based solely on latitude, there have been recent attempts at climatic classification which are based on temperature and precipitation, although the actual delineation of the individual climatic regions has been undertaken from many different standpoints.

GEOGRAPHICAL DEFINITIONS OF CLIMATE

A. Supan (1884) pushed the geographical viewpoint to the fore, by asking which areas possessed a more or less similar type of climate, and presented us first with 34 climatic provinces, and later with 35, which were primarily to be regarded as geographical unities. In fact, they differ only very slightly from the natural regions of Herbertson (1905), who divided the surface with regard to climate and surface relief. R. Hult (1892–3) has further stressed the climatological viewpoint and in a little-noticed work (noted by Ward, 1906) distinguishes 33 climatic zones which are delineated primarily by temperature conditions and secondly by precipitation and wind conditions. Thus he arrived at nine major zones, which he then proceeded to subdivide further into 33 zones which he subdivided again into provinces. The climatological viewpoint was put even more strongly by W. Köppen (1901). His very remarkable attempt at a classification of climate proposes a sharp division of climatic provinces on the basis of temperature and precipitation conditions, and both serve in delineating the boundaries. Biogeographic facts determine his seemingly arbitrary selection.

On the ground it seems possible to use climate (that is, the interaction of all atmospheric conditions) as a basis, for it imprints itself so clearly on the landscape that it becomes possible to distinguish whole climatic regions without having to start from long columns of meteorological observations. The influence of the climate on the character of the land surface is above all dependent on what form the precipitation takes. Whether it ultimately takes the form of rivers or glaciers is entirely dependent on the climate. A. Woeikof (1885–7) has emphatically stated that rivers are products of climate. From the climatological point of view it is important to know whether or not precipitation is completely evaporated thus leaving the land without water.

DIVISION OF CLIMATIC PROVINCES

Three different principal climatic provinces or regions may be distinguished on the earth's surface.

(1) The humid climates, in which more precipitation (N) falls than is removed through evaporation (V), so that a surplus in the form of rivers (F) runs off.

(2) The nival climates, in which snowfall (S) exceeds ablation (A), so that a removal in the form of glaciers (G) must ensue.

(3) The arid climates, in which evaporation absorbs all precipitation and could absorb even more, thus preventing a flow of river water.

We can characterise these three climates with the following formulae

$$(1)\ N - V = F > 0 \qquad (2)\ S - A = G > 0 \qquad (3)\ N - V < 0$$

Our three main provinces are separated by two important boundaries; the first is characterised by a balance between evaporation and precipitation, the other by one between snowfall and ablation. This latter boundary is the well-known snowline (SG), which can be expressed as $S = A$. Its other boundary has been termed the *dry boundary* (TG): $N = V$ is valid for this.

The snowline

The snowline has long aroused interest. It separates the areas under constant snow cover from areas occasionally under snow; this concept, therefore, often occurs in the context of landform evolution. Similarly, there have recently been probing discussions on the exact position of its location and, indeed, doubt has been expressed as to its very existence. In fact, its position is not constant; it changes from year to year, according to variations in ablation and snowfall, but in the course of the year it oscillates around a definite mean position. This is in no way related to a particular isohyet, for the snowline can be found in one and the same area at greatly differing heights, depending on exposure and surface relief, which in one place may be favourable to snow accumulation and in another hinder it. It has therefore become necessary to introduce an ideal height for the climatic snowline of a certain area, instead of local, observed heights. It is the height at which the snow that has accumulated in one year on a horizontal surface exceeds the total ablation. This value is important in comparing the snowlines of various regions; but in the delineation of the nival and humid areas the local snowline plays a considerable role. Under local conditions the same is true for the local snowline as for the climatic snowline on a horizontal surface, namely that above it more snow falls than can be melted in any series of years. Its position is therefore determined by the total snowfall and by the sum of the temperatures above 0°. A diurnal maximum of over 0°, which will melt the surface of the snow, will not decrease the snow cover as long as the meltwater freezes again during the night. Only a continuous period of warmth will *reduce* the snow. Therefore Finsterwalder and Schunck (1887) regarded ablation as more or less proportional to the snow-free period and the average temperature above freezing point during this time, and Kurowski made them proportional to the duration and average temperature of the period above 0°. But the only serious attempt to determine mathematically the relationship between snowfall, average temperature and length of frost-free period gave widely differing results from neighbouring glaciers (Machatschek, 1899) and today we are still a long way from understanding the individual elements of climate which determine the position of the snowline. It is the product of various factors, which are not yet fully understood.

The dry boundary

Less striking than the snowline is the dry boundary of the earth. Towards this boundary the humid regions become progressively poorer in rivers, so that these tend to disappear as the arid region is reached. It is evident that the position of this zone is noticeably influenced by the nature of the ground, just as the snowline is influenced by exposure; rivers will disappear on permeable sooner than on impermeable ground. It follows that arid areas are never completely without rivers; every heavy downpour is accompanied by surface runoff, although this is not a regular occurrence. Here we are concerned with torrents, rather than genuine rivers. Furthermore, numerous rivers flow out of the humid areas into arid regions. Whilst, however, they gradually increase in size in these humid areas, they progressively decline in the arid regions. Although the channels in arid areas are fundamentally different from those in the humid areas it is not always easy to separate the two. However, this difficulty does not prevent us from realising that the dry boundary is one of the most important boundaries on the earth's surface. The determination of its position has not yet been achieved, for we are not sufficiently acquainted with the controlling climatic factors. But a few bases for such a determination are available from an investigation of the relative conditions of precipitation (N) and runoff (F) in humid regions. They show that this relationship is not, as was formerly supposed, characterised in a river by a certain runoff factor, but can approximately be expressed by the following formula

$$F = (N - NO)x$$

where NO represents a critical precipitation value which may vary only slightly for neighbouring rivers and x is a proper fraction. To extrapolate, lack of runoff in the catchment area in question appears when

$$N = NO$$

The value NO therefore indicates the precipitation below which runoff ceases in humid areas. Axel Wallén in his work on central Sweden puts it at 100 mm—the values for Central Europe are around 420–30 mm. Merz (1906) has deduced the value 1100 mm for Central American rivers. As the average yearly temperatures for these areas are 1°, 7° and 24° respectively, it can be seen that the level of precipitation at which the runoff equals 0 increases with the temperature. But we are still a long way away from determining more exactly all individual elements which determine the position of the dry boundary.

PHREATIC ZONES

There are two main zones which can be distinguished in humid climates. In the first zone the precipitation can percolate into the ground and, depending on the ground's permeability, more or less fill it to form ground-water. This is not possible in the other zone because the ground is frozen. Here in the polar climatic province we have ground-ice instead of ground-water in what may be called the *phreatic* zone. The boundary of this ground-ice has aroused repeated interest. Fritz (1878) has represented it on a map which has been often reproduced; it coincides approximately with Wild's (1881) average annual temperature of −2°. In the polar climate sources of ground-water as well as ground-water itself are lacking. There is only some surface water which in summer succeeds in percolating through the thawing surface layer; it can move readily along such a surface and numerous movements of a purely superficial nature occur. In this sliding and partly flowing earth layer weathering is mechanical—with regular refreezing the water occasionally present in the surface layers shatters the uppermost rock layers and loosens them. The river is fed mainly by snow-melt and this generally produces considerable quantities of water within a relatively short period of time; a short high water stage in summer and a protracted period of winter low water characterise the polar rivers. The snow cover extends over the land for months, but not long enough to prevent tree growth. It is now known that tree growth is not related to the limit of frozen ground, as was originally supposed.

In climatic provinces characterised by phreatic zones, a greater or lesser part of the precipitation (depending on permeability) soaks into the ground and only joins the river after passage underground; these rivers then are only partly fed by the falling rain. The percolating water loosens the rocks along its path and attacks susceptible rocks with carbonic acid (H_2CO_3): it extends to the upper layer of the regolith and forms the characteristic leached or eluvial layer.

Provinces in phreatic regions

Within such a phreatic region we can distinguish individual provinces based on precipitation distribution. If rainfall is regular throughout the year, then feeding the rivers through precipitation or indirectly by ground-water continues regularly the whole year round and the rivers maintain a fairly constant flow. If, however, the precipitation shows an irregular distribution, exhibiting a clear distinction between rainy and dry seasons, then the rivers show a marked period of high water separated by periods of low water, and even of periodically dry stream beds. Such *occasional* channels are termed *wadis*. These occur at the boundaries of the arid and humid regions. In dry periods, percolation and ground-water supply ceases, and arid conditions develop which interrupt the humid conditions. Areas in which arid and humid conditions alternate annually are termed *semi-arid provinces.*

Towards the polar or nival province the phreatic region exhibits its special feature of the development of a regular snow cover, which can prevent the flow of water seasonally; then, after melting it can augment not only the ground-water but also the rivers. Accordingly, these rivers show a characteristic high level, which occurs according to the lateness of the melting; it occurs later in mountain rivers than in those of the plains and later in the season with greater proximity to the pole. This high meltwater stage often directly follows a period during which the streams are covered with ice. This subnival climatic province is separated from the nival province on its poleward margin by the snowline, and from the polar province by the occurrence of frozen ground. Its equatorward boundary is drawn where occasional snow cover ceases to contribute to the régime of the river. This occurs approximately where there is snow cover for about one month each year; where it lasts for a shorter period it causes no noticeable increase in precipitation storage. The subnival province, therefore, extends neither as far as the regions of permanent snow cover nor as far as the equatorial limit of snow fall, a point which has been closely examined by Hans Fischer (1887). Its boundaries still remain to be determined locally; and in attempting this, similar uncertainties will have to be met as with determining the dry boundary. The limits of the subnival province correspond approximately to the 1°C to 2°C isotherm for the coldest month. In the subnival province, as in the polar province, we can distinguish two subprovinces according to duration of snow cover; one with a predominantly snow-free period and the other snow-covered for most of the time. The boundaries of these two subprovinces coincide approximately with the treeline, and we therefore distinguish both in the polar province and also in the subnival province between forested and unforested regions.

The phreatic regions with a regular distribution of rainfall are typical of the humid climatic province. The latter breaks down spatially into subprovinces generally separated from each other by semi-humid or arid regions, namely the equatorial region with its high temperatures and abundant rainfall all the year round and the temperate region with its considerable annual range of temperatures. Neither prolonged ice formation on the rivers nor the regular appearance of a snow cover occurs in the latter region, although neither frost nor snowfalls are completely lacking. In the temperate humid subprovince, as in the subtropical province, the rivers usually exhibit high water in the cooler part of the year which, however, does not necessarily coincide with the period of most rainfall. During this time, evaporation is at its lowest and consequently the flow during this period is not only relatively, but also often absolutely, at its greatest.

PROVINCES IN ARID REGIONS

Just as the humid region can be divided into subregions in which precipitation percolates into the ground all the year round and those in which this process is temporarily or completely interrupted, so the arid region can be divided into two provinces in which aridity is important for all or part of the year. As we have seen, precipitation is never completely lacking in arid regions; it is always present, but perhaps not in sufficient quantity to be able to supply regularly flowing rivers. However, it can be of considerable importance in the development of torrents and quite a considerable vegetation growth, adapted to the dry climate. In this semi-arid climatic province the rainfall of isolated downpours often partially percolates into the ground, but cannot collect as extensive ground-water, because during the dry season it evaporates. Then it is brought to the surface again by capillary action. The 'percolating water' therefore, has no regular downward passage, as in the phreatic regions, and while it returns to the surface and is evaporated, it leaves behind those substances which it has dissolved at depth. Correspondingly, the leaching of the soil which occurs in phreatic areas is absent and is replaced by deposition of soluble salts, particularly calcium carbonate, in the upper layers of the soil. It is this calcium carbonate which so often makes up the hard surface crusts which are very characteristic of semi-arid regions.

In the completely arid province this rising and percolation of ground-water ceases and, consequently, no hard crusts are formed. The rocky surface is subjected to mechanical weathering only, as it has neither the vegetation cover of the humid regions nor the hard crusts of the semi-arid regions as protection against the wind. Hence the wind plays an extremely important role—eroding here, depositing there. According to temperature conditions the completely arid climatic province may be divided into two subprovinces: a temperate arid region with marked *seasonal* variation of temperature and a subtropical subprovince with a large *diurnal* range. A similar classification is possible for the semi-arid province.

NIVAL REGIONS

The nival region is characterised by accumulating snow deposits, both in the completely nival province where precipitation takes the form solely of snow and the semi-nival province where this is interrupted by rainfall. This rainfall, however, does not contribute to a decrease in the amount of snow; it causes moist surface conditions, which in turn favours a compaction of the snow and with the return of frost this again turns into ice. These hard crusts on the snow cover of our highlands play a particularly important role, but they may also develop in the completely nival regions as a consequence of intense insolation whereby the surface snow melts and the meltwater freezes again at very shallow depths.

In the nival region the land surface is protected from atmospheric weathering. But it is not improbable that a unique weathering process is initiated under the load of the snow and ice cover—and our attention has been drawn to this by Blümcke and Finsterwalder (1890). If a local increase in pressure causes a local fluidising effect on the base of the ice, then a moisturising of the basement rock may result. However, as soon as freezing sets in again, then this freezes too, and this can cause quite considerable frost-shattering forming fine dust particles. But this subglacial weathering of the ground remains far less important than the direct mechanical action of the glacier ice.

The ice erodes the surface and then deposits the eroded material when continual ground melting occurs—whether in parts of the basal ice where movement is minimal (Penck and Brückner, 1909) so that geothermal heat results in the liberation of englacial deposits, or at the periphery where the ice melts marginally.

The glaciers which originate in the nival region usually move out of this region and extend far into the subnival and polar climatic provinces, where melting occurs. The effect of glaciers extends far beyond the limits of the nival regions and the limits of former glaciation do not coincide with the extent of the earlier nival region. Just as the lakes of the glaciers extend out of the nival region, so the rivers of the humid region extend into the arid; therefore, the presence of typical fluvial activity at any one point does not indicate either that the area was formerly or is now a part of the humid region. Rivers entering the arid zone react in the same manner as glaciers entering the humid areas; they are consumed, they lose their water content—partly by direct surface evaporation and partly to the ground from which they gain no supply from ground-water and to which they lose a lot of water through filtering. In every respect they appear as strangers in the climatic province in which they find themselves.

KARST AREAS

If the presence of regularly flowing rivers is not indicative of a humid régime, then conversely the lack of rivers does not necessarily characterise an arid area. There are areas in the humid regions where the permeable nature of the ground not only favours the percolation of rainwater, but also the complete disappearance of whole streams. The karst areas are a good example of this. Numerous further examples can be found on extensive gravel sheets (*schotter*) and sandy landscapes, which soak up both rainwater and rivers. These pseudo-arid regions are distinguishable from the genuinely arid regions by the fact that the lack of surface water is combined with the occurrence of a good supply of water at depth, which succeeds in supplying springs. This spring-feeding water is lacking in truly arid regions: these have only filtered water which often extends far beyond the limit of surface water in allochthonous river beds.

Thus it is not one single feature which characterises a climatic region: rather, the character of a region is the sum of all its parts and it is possible to separate the individual regions by direct observation of these characteristics. This observation is valid for all the provinces discussed here.

REFERENCES

BLÜMCKE, A., and FINSTERWALDER, R. (1890). Zur Frage der Gletschererosion. *Sitzungsber. d. math.-phys. Klasse d. Kgl. Bayer, Adak. d. Wiss.*, **20**, 435

DE MARTONNE, E. (1909). *Traité de géographie physique*. Paris, 205

FINSTERWALDER, R., and SCHUNCK, W. (1887). Der Suldenferner. *Zeitschr. des Deutschen und Osterreichischen Alpenvereins*, 70

FISCHER, H. (1887). Die Aquatorialgrenze des Schneefalls. *Mitteilungen des Vereins für Erdkunde*, Leipzig, 97

FRITZ (1878). *Petermanns Geographische Mitteilungen*, table 18

HERBERTSON, A. J. (1905). The major natural regions. *Geographical Journal*, **1**, 300

HULT, R. (1892–3). Jordens Klimatomraden. Forsok till en indelning af jordytan efter klimatiska grunder. *Vetenskapliga Meddelanden af Geografiska Foreningen i Finland*, **1**, 140

KÖPPEN, W. (1901). Versuch einer Klassifikation der Klimate vorzugsweise nach ihren Beziehungen zur Pflanzenwelt. *Geographische Zeitschrift*, **6**, 593

KUROWSKI, P. Die Höhe der Schneegrenze. *Geogr. Abh.* **5**, 1, 115

MACHATSCHEK, F. (1899). Zur Klimatologie der Gletscherregion der Sonnblickgruppe VIII. *Jahresbericht des Sonnblickvereins für 1899*, 24

MERZ, A. (1906). Beiträge zur Klimatologie und Hydrographie Mittelamerikas. *Mitteilungen des Vereins für Erdkunde Leipzig für 1906*

PENCK, A., and BRÜCKNER, E. (1909). *Die Alpen im Eiszeitalter*, 951

SUPAN, A. (1884, 1908). *Grundzüge der physischen Erdkunde*. Leipzig, 1st ed., 129; 4th ed., 227

WARD, R. DE C. (1906). The classification of climate II. *Bulletin of the American Geographical Society*, **38**

WILD (1881). *Die Temperaturverhältnisse des Russischen Reiches*. St Petersburg, 348

WOEIKOF, A. (1885). Flüsse und Landseen als Produkte des Klimas. *Zeitschrift der Gesellschaft für Erdkunde*. Berlin, 92

—— (1887) *Die Klimate der Erde*. Jena, 39

3 Climate: Factor of Relief

EMMANUEL de MARTONNE

THE idea that surface relief is explained entirely by geology has become commonplace. Specialists know however, or they ought to know, that relief depends on climate almost as much as on the bedrock. It is the object of this article to recall some significant facts on this question and to try to characterise the nature of this influence.

Prolonged experience is not necessary to the recognition of the intimate relationship which unites the study of surface relief with that of study of the bedrock. Advances in geomorphology have always been bound up with those in geology: the first geomorphologists were geologists, such as Richthofen, Heim, Gilbert and Lapparent. One only has to open one's eyes to see at every step the primary role that geology plays in the explanation of relief forms. In teaching, it is difficult to keep pupils from the quite natural but extreme assumption that geology provides the total explanation. It would be far otherwise if one could only travel great distances and carry them quickly through different climatic regions.

The more the geographer travels the greater is he impressed by climate as a decisive factor in the geomorphology. It would be easy for me to invoke my own personal experience to support this point; but it seems preferable to use the observations of more renowned geographers, almost all of whom were geologists by their initial training. All have been led inevitably toward climatic explanations of surface forms.

There is no more significant example than that of Richthofen. The effects of climate made such an impression on him that they appear on every single page in his great work entitled *China*. None was more important than his insistence on the geographical contrast between the internal drainage regions of Central Asia and the peripheral regions with a more humid climate and a more developed relief. Herein lies the explanation of the enormous accumulations of yellow earth so characteristic of northern China to which he has given the classic name of *loess* which is applied in the Rhine plain to a similar silt.

THEORIES OF AMERICAN GEOGRAPHERS

The great extent of the United States and the climatic variety which is found there, from the glacier margins of Canada to the hot lands

of Mexico, has allowed American geographers, all originally geologists, to acquire very quickly an experience as fertile as that gleaned
by Richthofen on his journeys in Asia. The particular originality of
theoretical writers and of the classic descriptions given by Powell,
Gilbert or Davis is rooted in the attention they gave to different
conditions of erosion in humid and dry regions, forested and grassland regions, and oceanic and continental environments. It was in the
exploration of the Rocky Mountains and the territories of the west
that the masters of geomorphological method perfected their approach
and enlarged their conceptions. One understands even more graphically the value of such contrasts when one is rapidly transported
by powerful transcontinental express trains from the verdant ridges
of New England and Pennsylvania to the badlands of the west,
intensely gullied by stream action on the bare earth; or from the deep
valleys, with their wooded regular slopes which cut into the Allegheny
Mountains, to the tremendous canyons of the Colorado and the
Arizona with their vertical rock sides in brilliant colours, and their
gigantic talus slopes. It is impossible to escape the impression that
the sculpture of the relief if not due to different tools has been exe-
cuted in an altogether different style in both the humid Atlantic
zone and the dry regions of the west. Gilbert has translated this impression very clearly in the form of a geomorphological law which
states that, under identical geological conditions, slopes are steeper
in a dry climate than in a humid one.

It is interesting to discover exactly the same law coming from the
pen of General de la Noë and Emmanuel de Margerie in *Les Formes
du Terrain*. The experience of de la Noë did not extend outside French
territory; the appearance of the Mediterranean region of Provence
and particularly of Algeria with its high, dry plateaus, sufficed to lead
him quite independently of Gilbert to the same law expressed almost
exactly in the same terms.

Since the time of Richthofen, Gilbert and de la Noë, geomorphology has made great progress. It would seem that the moment has
arrived when one would scarcely hope to discover new interpretations of the normal erosional system such as may be seen in the
countries of the temperate zone. But new horizons open before those
who embark on the study of the more exotic regions where a poorly
understood suite of processes prevails in extreme conditions of
temperature and humidity. The development of a scientific knowledge
of desert and polar regions poses new problems and allows us to

glimpse new solutions even for the interpretation of relief forms in the temperate zone, given climatic changes.

DESERTS

Fifty years ago we knew next to nothing about deserts. The Sahara was the 'great sand sea', ancient bed of the ocean comparable in scale to the moving sands on the margins of the sea. The travels and scientific explorations of Barth, Rohlfs, and Foureau have corrected this erroneous idea in exposing the great richness of arid forms. The exploration of the Kalahari desert in South Africa by the geologist Passarge, the Central Asiatic deserts by that indefatigable pioneer Sven Hedin, the Australian desert by Horn and the comparable studies by J. Walther, have revealed the ensemble of forms specific to the deserts and whose development appears to be due to the action of forces unknown elsewhere: wind, very infrequent but torrential rains, and the overheating of rocks. In addition to the sand desert, with its various types of dunes, we have the rock desert, the *hammada* of the Sahara, a plateau covered with rocks polished and faceted by the wind; the *gours*, erosion residuals fretted by the wind until they take the form of mushrooms; blind valleys, the rocky anastomosing gullies to which the name *chebka* is given in the Sahara; the deserts of clay and silt, termed *yardangs* in Central Asia, sculptured by the wind into parallel crests.

Set on this course from which so many new horizons reveal themselves, it is natural that researchers have sometimes allowed themselves to get carried away. It seems certain that in the category of desert landforms many features have been included which develop in fairly humid climates but that they were too readily explained in terms of the nature of the rocks; the role of wind as an agent of erosion has certainly been exaggerated by J. Walther. Forms similar to those of the African or American deserts are to be found in Europe: the honeycombed crusts of weathered sandstone to be found in Swiss Saxony have given rise to much discussion in the specialist German literature; some workers would have them as proof of the existence of a desert climate during the Quaternary, while others explain them by simple decomposition of the sandstone in a humid climate. The geologist Passarge has been led by his studies in South Africa to evolve an attractive theory, according to which the wind and rill action under desert conditions is said to have effected widespread landscape reduction over almost the whole of the African continent. Study of regions of normal erosion has taught us to recognise ancient erosion surfaces in mountainous regions whose elaboration is readily explained by the action of fluvial erosion carrying its work on to the ultimate degree to what Davis has termed the *peneplain*. If one is to accept the theory of Passarge, interpretations based on the latter conception ought to be revised if examination reveals that the former climate of the region studied could not have been arid.

POLAR REGIONS

While new perspectives have stimulated the scientific exploration of the deserts, those of the polar regions have not been less fertile. The study of Greenland and Iceland has helped us to understand geographical conditions which prevailed during the Quaternary, when vast ice sheets covered Europe and America. These colossal glaciers profoundly modified the surface forms, obeying the same laws as those of the present polar glaciers.

Beyond the ice-covered regions, the cold regions provide us with a collection of forms due to very particular environmental conditions controlled by climate, and a beginning has been made on the study of these in the last few years. In Spitzbergen, Labrador and Alaska, block fields are widespread and are due to the same phenomenon which produces the chaos on certain Alpine summits: the shattering of rocks under the influence of water expansion which freezes in the cracks it is able to penetrate. But the most curious feature to see is the grouping of the largest stones into ridges which describe circles or linked polygons. Such highly distinctive regolith is due to the alternation of freeze and thaw, so frequent in Spitzbergen, where it has been described by the term *polygonal soil*. On the more steeply inclined slopes the effects of freeze-thaw manifest themselves in another form: the superficial debris, made up of clay, sand or fragments of argillaceous rocks, is constantly moving down like a plastic mass. Parallel ridges, generally crescentic, betray this movement and recall in retrospect the appearance of glacial moraines. In Alaska and in some parts of the Rocky Mountains, veritable rock rivers can be seen running down the face of mountain scarps. The American geologist R. Capps has described these by the very expressive and appropriate term *rock glaciers*. In fact, some of these rock glaciers have their origin in true cirques and overrun the forest as they pour down the steep slopes. Their speed of movement is comparable to that of some Alpine glaciers and is facilitated by snow meltwater which infiltrates into the mass and so is subject to alternating freeze and thaw.

TEMPERATE REGIONS

The moment would appear to be at hand when the experience acquired in the study of forms specific to present-day Arctic regions will help us to understand certain peculiarities of the temperate regions which, during the glacial periods of the Quaternary, must have suffered a climate similar to that of Labrador or Spitzbergen at present. The great spreads of granitic or sandstone blocks known by the general term *mers de rochers* or *felsenmeere*, so common in the mountains of Central Europe, can scarcely have formed in the present climatic conditions. Lozinski remarks that they are only found in the mountains in the marginal zones of the former Pleistocene ice-sheets which moved south from Scandinavia and covered all the northern part of Central Europe; this would constitute, according to his own expression, a *periglacial facies* in the decomposition of the massive rocks. The morainic nature of the deposits which clutter the high mountain valleys of Central Europe and the Carpathians has been discussed many times. There is little reason to doubt that in certain cases these are *felsenmeere* hardly reworked. In other cases it is probable that the material is that of ancient rock glaciers comparable to those found in Alaska and derived from little cirque glaciers. This hypothesis, which was suggested to me by the work of Capps appears to correspond to reality, for I have seen in the Doessen valley in the Tauern its tiny, rapidly disappearing glacier reduced to a névé, its lower extremities becoming a true rock glacier and descending until it reaches the level of Lake Doessen. What a succession of hesitations might I have experienced in my research in the southern Carpathians had the rock glaciers of Alaska been unknown to me.

These examples will serve to show the importance that consideration of climate takes and will in the future take in the explanation of surface relief. Do they allow us even to guess what is the principle of this imprint by the climatic forces in terms of landscape forms? If observation reveals more and more different forms under different climates, then these forms are the result of sculpturing and modelling by the atmospheric agents. Let us endeavour to examine precisely how the several climatic elements can influence the course and thus the style of this process.

TEMPERATURE AND HUMIDITY AS EROSION FACTORS

Streams are the best known agents of erosion. One is less impressed by the slow but continual activity of rock decomposition whose status and activity are regulated by the two principal elements of climate: temperature and humidity. Erosion, properly speaking, works in leaps and bounds; decomposition goes on everywhere without interruption.

High atmospheric humidities, generally associated with abundant rainfall, favours chemical decomposition by the action of water infiltrating below the surface. In areas of humid climate, the weathered mantle is generally deep, outcrops of bedrock are rare even on steep slopes and in areas of resistant rock; the products of decomposition form an almost continuous mantle which masks the irregularities beneath the solum and softens all the landforms. In areas of dry climate, chemical decomposition is less active while mechanical decomposition, due particularly to variations in temperature, is all the more in evidence. The debris is coarser and cannot remain in place on the slopes; it rolls down leaving the rocky escarpments bare, forming at their foot a series of talus slopes.

Here we come close to the reasons for the contrast between the sharper landforms of dry regions and the gentler forms to be found under humid climates. This contrast is familiar enough as witnessed by the differences between the limestone plateaus of the Paris Basin and those of Provence. The observer used to the southern parts of the country searches in vain on the pleasant valley sides of the Bourgogne, or along the continuous escarpments of Lorraine and is surprised not to see more outcrops of bare rock. The contrary impression awaits the geographer coming from the north who, in the south, is naturally disposed to attribute to the rock a much greater resistance when he sees it outcropping everywhere.

The contrast is even more striking in the United States between the humid Atlantic region and the far west. A journey of a few weeks in the strange and marvellous world of the deserts and mountain steppes which extend between the Rocky Mountains and the Pacific suffices to accustom the observer to the sharp, clear-cut landforms;

we are used to reading the structure of an area in the topographic forms, the slightest resistant bed expressing itself in an escarpment that is easily followed. The gentle forms of the Appalachians are a deception for the geologist. The highest summits in Carolina are rounded domes; rocks as hard as the Cambrian quartzitic conglomerate, or the Knox dolomite produce only discontinuous escarpments similar to those of the Jurassic limestones of Lorraine or the crags that outcrop on the slopes of some incised valleys in Brittany. The explanation for this is made quite clear by exposures in railway cuttings or recent roads where enormous thicknesses of deeply altered red soil are found together with granitic gravel; these exposures sometimes exceed 30 m in depth without reaching bedrock. With its humid eastern coast and its relatively dry interior plateaus, Brazil provides similar contrasts. According to Branner, the granitic sands of the coastal area reach a thickness of almost 100 m. In the interior the soil is much less thick and the rocky escarpments of the mesas are characteristic.

Humidity is not the only thing to consider in attempting to understand conditions of rock decomposition and its results. The average values and particularly the variations of temperature are also very important. The equatorial climates with their constantly high humidities are areas where chemical decomposition is the most active, soils are deepest and rocky escarpments the most rare. Every geological explorer has complained about making journeys without finding a single outcrop. In the temperate regions there is a fairly clear difference between oceanic and continental climates. The former are at once more humid and more equable; soils are generally rather deeper. The latter are drier and more extreme, temperature variations being fairly severe so that mechanical decomposition is of almost equal importance with chemical decomposition.

DESQUAMATION

In cold climates, mechanical decomposition predominates to the same extent as in the desert regions; it is particularly the action of freezing, causing the rock to fracture along fissures which the water has penetrated, which creates the masses of coarse debris so characteristic of polar regions and the summits of high mountains. The sharp forms of the Alpine mountain tops are due for the most part to the effects of such weathering cutting into the bare rock. The term *desquamation* has been given to a form of mechanical decomposition which, especially in granitic rocks, gives rise to the strangest features. The word is very expressive for these are like splinters which have detached themselves from the surface of the rock. They result in the formation of rounded peaks rather like cupolas or sugarloaves. Such are the famous Corcovado and the Sugarloaf whose fantastic silhouettes tell the navigator from some distance that he is approaching Rio de Janeiro; such also is the Half Dome of the Yosemite of California. On the flanks of the incised valleys, the sheets of rock detach themselves in a form parallel to the slopes which thus form vast smooth convex surfaces.

The distribution of these forms is quite surprising: domes and rounded peaks are found in the Sudan, Brazil, California and in some parts of the Rocky Mountains; desquamation of granite surfaces is to be seen in some Alpine valleys, notably in the valley of the Aar, where it is as well developed as on the sides of the Yosemite Canyon in California, and also on the high plateaus of Madagascar. What is there in common between these regions with such different climates? The answer would seem to be in the intensity of solar radiation and the alternation of dry and wet periods. The rarified air of high mountain areas allows direct solar heating quite as much as the pure dry air in California, the Sudan or Brazil during dry periods. It is heating of the rock surface which causes the shattering of rock fragments especially when there is severe cooling brought on by rain showers. A recent observation from Madagascar by Carrier illustrates this well: on the high plateaus of the Emyrne, the local inhabitants work the granite for tombstones and, in so doing, help the work of nature by lighting fires against the rock surfaces and then dousing them with cold water.

CLIMATE AND EROSION AGENTS

If it is worthwhile to stress the variety arising from different conditions of weathering according to climate, there is scarcely any need to stress the fact that climate controls the distribution of the agents of erosion: glaciers and glacial erosion, rivers and fluvial erosion, deserts and aeolian erosion.

Regions in which snowfall dominates the precipitation and in which temperature is too low for the sun to remove it completely have a very distinctive relief: the accumulation of snow, transformed into ice, tends to bury the relief and to swamp all the forms; however, where the ice runs off by gradual melting, the relief is moulded irregularly by glacial erosion, combined with very active mechanical weathering of the rocks above it. Thus it is that Greenland, with an altitude of more than 3000 metres, has the appearance of a remarkably uniform plateau, only the margins being accidented thanks to the alternation of rock spurs and glacier tongues. In the Alps themselves, the highest summits such as Mont Blanc and Mont Rose have the appearance of domes capped by ice: the needle-like peaks and sharp *arêtes* occur at lower altitudes where glaciers melt as they reach below the limit of perpetual snow.

Beyond the zone of ice-melt in an equable climate, the landscape still bears the imprint of glacial action because the debris transported by the ice makes up enormous accumulations called moraines, and the torrents supplied by meltwaters from the same glacier rework and distribute the same debris over a wide area. Thus the morphological action of glaciers makes itself felt well beyond the limits of the glacial climate.

Aeolian effects are no more limited to deserts. They make themselves felt everywhere where a vegetal cover is lacking. If they are found to be dominant in the driest terrains such as the Sahara and the Asiatic or Australian deserts that is because the vegetation is practically entirely absent and the regolith, made of fine debris, is robbed of all cohesion as water is lacking at the surface. Sand, transported by the wind, and sometimes even the fine gravel which may rise to head height, are the instruments by which erosion can attack even solid rock. But the distinctive topographical characteristics of deserts are due perhaps less to the work of the wind than to mechanical weathering and the accumulation of debris which this produces, no running water being capable of transporting it to the sea.

Fluvial erosion

The influence of fluvial erosion extends over the largest proportion of the continents but its characteristic aspects are so variable that one is led in reality to distinguish several provinces as different in their relief as in their vegetation and their climate. Erosion by torrents, which has generally been considered specific to mountain areas, is in fact a special type of fluvial erosion found in areas where the climate is such that discharge is as irregular and as violent as in mountainous regions. On the margins of almost all the deserts of the world can be seen mazes of sharp ravines separated one from another by delicate crests. When these were encountered by the French trappers on the great plains at the foot of the Rocky Mountains they gave them the name *mauvaises terres*. This has since been translated by the Americans as *badlands* and extended to many regions of the same type. There are comparable badlands in almost all of the steppe areas, for example, in the Argentine Pampas, where the name *barrancos* is given to steeply sided torrents; in southern Russia where deforestation encourages the gullying of valley sides; in Turkestan; in the loess country of China and even, if one can judge from the photographs of the Bailey Willis expedition, in the granitic mountains of the Petchili. The essential condition for the development of badlands appears to be violent and infrequent rainfall with, however, sufficient discharge to maintain continuous flow in the trunk valleys and almost continuous flow in the secondary valleys.

Rain is not entirely absent in deserts; fluvial erosion is therefore associated with aeolian erosion, though runoff is even more irregular than in the badlands. In hilly regions where stream courses are fixed due to their incision into the rock, master streams are no longer distinguishable although it is often a matter of a labyrinth of anastomosing gullies; this is the nature of the *chebka* of the Sahara. On the plain lands stream beds shift constantly, almost at each shower; the alluvium thus becomes spread out over vast surfaces, making a *glacis* inclined with the gradient of a torrential bed. These strongly sloping alluvial plains border the desert mountains almost everywhere, in the Algerian Sahara, the basins of the west of Mexico, Iran or Central Asia. Their appearance always surprises the geographer conditioned to the gentle slopes of the alluvial plains of humid regions, and when they are broken up into terraces as they commonly are, one is tempted to invoke in explanation a degree of uplift when in fact they are produced only by a change of climate.

Erosion conditions in the past

In research on the evolution of topographical forms one must without doubt accord increasing importance to those hypotheses proposing that past erosion conditions were different from those of the present due to climatic conditions being different. Physical geography has been involved in this kind of relationship for something like half a century through the studies of glacial geologists. We know that the first men were witnesses of an extension of glacial climate incomparably greater than that of the present day. Two-thirds of North America and almost half of Europe were covered during the Quaternary epoch by ice sheets comparable to those of Greenland or the Antarctic continent. While we do not have a precise date, we can say that the number of years which have elapsed since the disappearance of these glaciers is to be reckoned only in thousands. As fluvial erosion has increased in response to the warmer climate of the present, it has only partially effaced the glacial imprint. It may be said that over more than half the area occupied by glacial ice during the Quaternary has a landscape which does not accord with present conditions of erosion. In almost all the great mountain chains—the Alps, the Pyrénées, the Caucasus, and the Himalayas—the valley forms cannot be explained in terms of the work of rivers and torrents that we see acting today. An understanding of these forms only begins when recourse is made to the action of glaciers which once extended to the gates of Lyons, Munich and Milan.

The study of deserts also owes a degree of progress to the consideration of climatic change. The extension of fluvial erosional forms in the Sahara, for example, is more evident from day to day and appears inexplicable in terms of the present climatic conditions. The general cooling which led to the extension of the glaciers in the temperate zone also made itself felt in the tropical zone where its reduced evaporation increased humidity and gave rise to more abundant and regular rainfall and allowed water to circulate through valley systems which are now always dry. Gautier and Chudeau interpreted the network of great wadis in the Sahara as the work of true rivers. Passarge also found in the Kalahari desert indications of a pluvial period in the Quaternary. While there is no reason to believe that these deserts once experienced a truly humid climate, they must have approached the conditions of steppe.

The study of the interior drainage basins of the western United States has revealed even more curious morphological facts which accord with recent climatic changes. Gilbert and Russell demonstrated that the Great Salt Lake and several smaller lakes in the neighbourhood are the last vestiges of immense sheets of water which, in the humid period of the Quaternary, covered a territory as large as France. The action of waves and currents has left on the

margins of these lakes beaches which are still perfectly recognisable.
Every detail of typical littoral forms such as may be seen on the
present coast are to be found there: terraces, lagoonal barriers,
deltas, among other features.

Beyond all these facts relative to the morphological effects of
changing climates, the mind is alerted to glimpses of new perspectives. Given a radical change in climatic conditions as a result of the
modification of the composition of the atmosphere, for example a
marked variation in the amount of carbonic acid, the conditions of
rock decomposition would be severely changed. Chemical decomposition would become more rapid in humid climates, and the reduction
of landforms more marked. Is it not easier to envisage a complete
planation under such conditions producing these uniform surfaces
called peneplains? Perhaps it is in such terms that one must look for
the explanation of these absolutely flat platforms which truncate at
several levels the folded limestone strata in almost all the karstic
regions, as much in Istria and Dalmatia as in Languedoc or the Jura.

THE INFLUENCE OF HUMIDITY ON THE
EVOLUTION OF RELIEF

The part which consideration of climatic change is called upon to
play in the study of relief forms is now apparent. If one asks what is
the element of climate whose variations appear to exercise the greatest
influence on the evolution of relief, the answer would seem to come
without hesitation: humidity. The amount and incidence of atmospheric precipitation determines the mode of erosion and even the
characteristics of rock decomposition. Temperature is a factor which
certainly cannot be neglected, but its action is for the most part indirect; low temperatures lead to a predominance of solid precipitation and a glacial régime; high temperatures stimulate evaporation
and tend to diminish the volume of streams and drying out the soil,
exposing it to the action of torrential streams or even, if the vegetation is not sufficiently fixed, to wind action.

It is perhaps possible to respond to an even more embarrassing
question: how may we estimate the relative importance of climatic
and geological considerations in the explanation of relief?

It is quite evident that the major relief forms of the earth are
determined by geological events such as folds, uplifts, downwarping,
and ancient erosional episodes. The distribution of land and sea,

plains and mountains is not a climatic fact: on the contrary, it conditions the climates. What is directly due to the climate is the landform detail, the sculpture or topographical modelling. To the varying conditions of climate the countryside owes not only the colour of the sky and the waters, and the variations in the richness and continuity of the vegetation cover, but the very forms of the valleys, ridges and mountains which make up the framework of all life and of all activity. It seems that we may speak of *topographic facies*, just as one speaks of geological, botanical or zoological facies; and it is agreed that the topographical facies forms the hall-mark that climate imprints on the relief and whose gross outlines have been affixed by geological development.

GEOGRAPHIC FACIES

It would be a geographical exercise of considerable interest to distinguish the nature and extent of the geographical facies on all the continental surfaces. I have attempted some work of this kind in establishing an outline of the distribution of the dominant forms of continental erosion in my *Traité de Géographie Physique*, p. 406. Some types have already been defined.

The warm humid facies

This is characterised by rock decomposition, especially chemical; great thicknesses of residual soils, even in the mountains; rarity of escarpments and rock outcrops; the gentleness of all slopes; the very low slope of the plains and large alluvial cones. The equatorial regions of Africa and America provide the greatest extension of this facies.

The humid temperate facies

This presents almost the same characteristics but in less marked form. The thickness of the residual soils is less, slopes are steeper. This prevails in temperate regions especially where the climate has no marked continental character, especially in western and central Europe and in the eastern United States.

The dry season facies

Again this is distinguished by its mechanical weathering; the debris mantle is of variable thickness and made up of coarser fragments.

Rocky escarpments are more frequent, gradients of slopes, river beds, plains and alluvial cones are much higher. Stream flow is irregular, and erosion which is more violent and spasmodic often has the character of torrential erosion and forms badlands whose extent varies with the general slope and the nature of the terrain. This facies is found in almost all steppe areas adjacent to deserts, in the tropical savanna regions and in the subtropical lands with a dry season such as the Mediterranean countries and similar areas (California, Cape of Good Hope for example).

The desert facies

This is one of the clearest facies and demands no extensive definition. Mechanical decomposition acts almost alone in producing very coarse and very unequally distributed debris. This debris accumulates in place and cannot be transported by water. The bare earth is attacked by the wind: the stony *hammada*, the *chebka* channels, the saline basins, the great sandy wadis and the dunes are the commonest desert landforms.

The cold dry facies

This is still little known, being a type of desert facies which prevails in polar regions and those parts of the high mountains not actually covered by glaciers. The air is very dry, the generally thin snow cover disappears for a part at least of the year; the soil is frozen to a great depth and thaws at the surface during the warm season, giving rise to curious forms which we refer to as *polygonal soils*. All rock outcrops are attacked by mechanical weathering particularly by freezing, and vast chaotic sheets of coarse debris are produced.

The cold humid facies (glacial facies)

This is as well-known as the desert facies. The masking of the landforms in areas of perpetual snow, their delicate chiselling out in the ablation zone of the glaciers, the enormous accumulations of debris in the outer zone with its even softer climate—these are the principal characteristics.

It is obvious that there are innumerable transitions between the different facies. This complication is increased by the fact that the landscape tends to preserve evidence of past climates which differ from the present. A complete study of the extent of topographic facies would reveal, it would seem, quite major complications; but what interesting problems would it not unearth? Without neglecting the geological side of the earth's morphology, one is led to hope that geographers will move more and more towards the study of the climatological aspects of relief development and towards the fertile analysis promised by topographical facies.

4 The Cycle of Glaciation

WILLIAM HERBERT HOBBS

THE CIRQUE AND ITS RECESSION

WITH the advance of knowledge about the sequence of conditions affecting glaciers, it has come to be generally recognised that for any given district the factor of supreme importance is temperature, a very moderate change in the average annual temperature being sufficient to transform a temperate district completely, and to furnish it with snowfields and mountain glaciers.

With the probability that such climatic changes would be initiated slowly, the first visual evidence of the changing condition within all districts of accentuated relief would probably be a longer persistence of winter snows in the more elevated tracts. These accumulations of snow would eventually contribute a remnant to those of the succeeding winter, and so bring on a cycle of glaciation. From this beginning the cycle is an advancing one until a culmination is reached corresponding to the most rigorous of the climatic conditions. A resumption of a more genial climate would result in a reverse series of changes, terminating early as the winter's fall of snow is insufficient to produce permanent snowdrifts even in the higher areas. It is therefore proper to speak of the *advancing* and the *receding hemicycles of glaciation*.

Mountain versus continental glaciers

The landforms which result from glaciation within districts of strong relief when not entirely submerged beneath snow and ice, are totally different from those which are sculptured beneath a glacier of continental dimensions.

So far as low-level mountain glaciation is concerned, the erosive processes are known to be identical with those of continental glaciers, namely, abrasion and plucking. The larger proportion of projecting rock-masses in the case of mountain glaciers will, however, presume a greater emphasis upon lateral undercutting from the operation of both these processes acting conjointly. It is the operation of an additional denuding process of the first importance, head-wall erosion, that differentiates all types of mountain glaciers from continental ones. This distinguishing process is responsible for the development of the cirque which is known by a variety of names in different glacier districts. In Scotland it has been generally referred to as the *corrie*, in Wales as the *cwm*, and in Scandinavia as the *botn* or *kjedel* (*kessel*). In the scientific literature of the subject, the Bavarian–Austrian word *kahr* has been used with increasing frequency for the same topographic feature. In view of this diversity in resultant topography, and despite their close genetic relationships, we would do well to separate sharply in our discussions continental glaciers from all other types, which latter we may include under the broad term of mountain glaciers.

Cirque erosion

It is safe to say that no topographic feature is more characteristic of the mountains which have been occupied by glaciers than is the cirque. Approaching a range from a considerable distance, there is certainly no form which so quickly forces itself upon the attention.

Up to the beginning of the twentieth century, however, few geologists had greatly concerned themselves with the erosion conditions at high levels, the work of Richter being on the whole the most comprehensive. The whole subject of cirque erosion was rather generally ignored, as it is indeed today.

The discovery of the method by which the glacier excavates its amphitheatre must be credited to a keen American topographer and geologist, Willard D. Johnson. In fact, to him and to another American topographer, François E. Matthes, we owe the most of what is known from observation concerning the initiation and the development of the glacier cirque. Reasoning that abrasion was incompetent to shape the amphitheatre, Johnson soon surmised that the great gaping crevasse which so generally parallels the cirque wall and is termed the *bergschrund*, went down to the rock beneath the névé, and that it was no accident that glaciated mountains alone 'abound in forms peculiarly favourable to snowdrift accumulation'.

Everywhere in the crevasse there was melting, and thin scales of ice could be removed from the seams in the rock. The bed of the glacier, elsewhere protected from frost-work, was here subjected to exceptionally rapid weathering. By keeping the rock wall continually wet, and by admitting the warm air from the surface during the day, diurnal changes of temperature here resulted in very appreciable mechanical effects, whereas above the névé only the seasonal effects were important.

These observations of Johnson (1904) contrast with the suppositions of Richter, who believed that the maximum sapping upon the cirque wall occurred above the surface of the névé. The function of the *bergschrund*, which separates the stationary from the moving snow and ice within the névé, is thus found to be of paramount importance in the shaping of the amphitheatre.

SCULPTURING OF THE UPLAND

The upland dissected

Having obtained a clear conception of the process of head-wall erosion through basal sapping, Johnson was in a position to account for the topography which he encountered in the High Sierras of California.

With little doubt the failure to recognise generally the importance of this process of cirque recession, clearly here a more effective agent than abrasion, is to be explained by the fact that in Europe generally, and in the Alps in particular, one looks in vain for evidences of the earlier and more significant stages of the process. Glaciation was so vigorous here as to cause the removal of all summit upland. Within the arid regions of the western United States, a more fruitful field for study is to be found. Here the work of Johnson has been supplemented by that of Gilbert (1904) and Matthes (1899–1900). Perhaps nowhere are the early stages of the process so clearly revealed as in the Bighorn Mountains of Wyoming (see Fig. 4.1).

Remnants of the preglacial surface will, in any given district, be large or small according to whether or not nourishment of the glaciers has been sufficient. The Uinta range, which extends in an east-west direction, and, like the Bighorn Mountains, has a core of homogeneous granitic rock, displays this fact. An examination of Atwood's map (1909) shows that to the eastward, where the precipitation has been least, the remnants of the original upland are more considerable. This qualifying condition of glacier nourishment will be subject to some modification because of peculiarities in snow-distribution.

Modification in the plan of the cirque as maturity is approached

Owing to the fact that the sapping process within the cirque operates on all sides, its early plan, when the upland surface is supplying snow from all directions, will approach the circle (see Fig. 4.1). Moreover, in this stage the cirque will be but little, if any, wider than the deepened and widened valley below. Later, with the continuation of the sapping process, the cirque becomes enlarged to such an extent

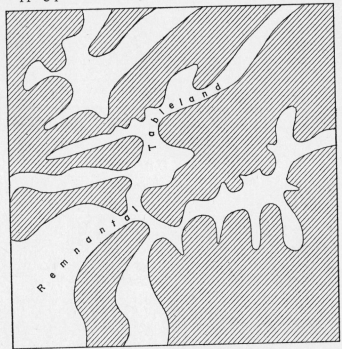

Fig. 4.1   Preglacial upland invaded by cirques, 'biscuit-cutting' effect, Bighorn Mountains, Wyoming

that its sides form recesses in the walls of the valley. Thus, in the plan, the glacial valley of this stage bears some resemblance to that of a nail with a large rounded head.

As the upland is still further dissected, the cirque becomes more irregular in outline and widens into a roughly elliptical form, in reality a composite made up of several cirques of a lower order of magnitude.

GROOVED AND FRETTED UPLANDS

An early product in which large remnants of the upland surface still remain, may well be designated a *grooved* or *channelled* surface. As the hemicycle advances, it will be observed that on the flanks of the range are found the largest remnants of the original upland surface (see Fig. 4.1) owing to the tendency of the cirque to push its side walls out beyond the limits of the U-shaped valley below. With complete dissection of the plateau no tabular remnants are to be discovered. The general level of the district has now been lowered, but above this irregular surface project one or more narrow pinnacle ridges, which at fairly regular intervals throw off lateral palisades with crests which fall away in altitude as they recede from the trunk ridge. In general terms, and describing the major features only, we have here a gently domed surface, on which is a fretwork of comb-like ridges projecting above it. This surface may be called a *fretted upland*. Such a condition is realised in the Alps, and is seen to special advantage from the summit of Mont Blanc.

The fretted upland differs from the grooved upland of an earlier stage of the cycle in the complete dissection of the surface. The character of the fretted surface is well brought out by the topography of the Lofoten Islands off the arctic coast of Norway, where the effect is somewhat heightened through the submergence and consequent obliteration of the irregularities in the floor.

Characteristic relief forms of the fretted upland

In the earlier stages of mountain glaciation the upland is channelled by valleys U-shaped in their upper stretches, and somewhat broadened into steep-walled amphitheatres at their heads. With the complete dissection of the upland, the coalescence of the many cirques at last cuts away every remnant of the original surface and yields relief-forms which are dependent mainly, as already stated, on the initial positions of the cirques.

If there be a highest area within the upland, the snow will be carried farthest from it by the wind, and this will be in consequence the last to succumb to the cirque-cutting process. The dome of Mont Blanc in the midst of a forest of pinnacles, no doubt owes its peculiar form to the fact that it dominated the preglacial upland. Elsewhere within the upland the coalescence of cirques has produced comb-like palisades of sharp rock-needles which have long constituted the *aiguille* type of mountain ridge. In the literature of physiography, such ridges have perhaps most frequently been designated by the term *arête* (fish-bone). I propose to use for all such palisades of needles derived by this process the name comb-ridge as the best English term available.

In every mountain district maturely dissected by glaciers, can be found sharp horns of larger base and especially of higher altitude than the individual minaret-like teeth of the comb-ridges. They are further in contrast with the latter by having an approximately pyramidal form, and a base most frequently a triangle with flatly incurving sides. They appear most frequently at the junction points of the comb-ridges between three or more important snowfields. Such forms are generally termed *horns*. The prominent horns of any glaciated mountain district no doubt occupy positions corresponding in the main to the more elevated areas in the original upland surface, since such positions would be the earliest cleared of snow, and hence the last attacked by the cirques. After complete dissection of the upland the comb-ridges which fret its surface will be attacked from opposite sides, and their crests will be first lowered at the points of tangency of the adjacent cirques, generally near the middle points of their curving outlines. The skyline of the ridge will thus be lowered in a beautiful curve forming a pass or col.

Though the sapping process at the base of cirque walls up to maturity is doubtless far more potent than abrasion and plucking upon the floor of the amphitheatre, it seems likely that in the subsequent stage the reverse is the case. This would at least explain the tendency of glacier valleys to deepen rapidly in the higher altitudes, or, in Johnson's phrase, to get 'down at the heel'.

MONUMENTED UPLANDS

In a visit to the Glacier National Park, the writer was impressed that there was a type of topography which indicates a still later stage of sculpture by mountain glaciers than does the fretted upland as exemplified by the Alps. The most striking peculiarities of this type are found in the unusual number of isolated sharp peaks of monumented aspect, and this is combined with a general absence of the comb-ridge and a frequency of unusually low cols or passes. Unlike the true horns of the Matterhorn type, which in the fretted upland are relatively few in number and may perhaps represent by their summits points near the original surface of the upland, the monuments of the northern Rocky Mountains show a tendency to appear in pairs, and in many instances at least they are remnants of lower portions of the preglacial surface.

Both in the Bighorn Range and in the Glacier National Park

the glaciers have today nearly or quite disappeared, being now represented by small horseshoe or cliff glacierets only. The earlier conditions of nourishment were, however, as we know from more or less extended studies, notably different from those of today. In the Bighorn Range the glaciers of Pleistocene time extended far down the valleys, where strong terminal moraines are found to mark the limits of their advance.

In the Glacier National Park district the Pleistocene glaciers occupied the entire valleys within the range and spread out eastward their aprons of piedmont type. We can use here the term *monumented* upland to describe the extreme type of mountain sculpture which is represented in the Glacier National Park and which is believed to be due to continued glacial action upon a fretted upland like that of the Alps. Cirque enlargement carried to this stage has sapped the main comb-ridge so as to largely obliterate the *aiguille* type of crest or *arête*. Matterhorns have in the process been reduced in size as the cols are progressively lowered and widened and are transformed into *arêtes*. The last remnants of the upland to be removed by this continued cirque enlargement are found away from the original divide and outward toward the flanks of the upland, for the reason that in their later stages cirques enlarge excessively on their lateral walls. A good illustration of this tendency is supplied by the gently sloping summit plane of Quadrant mountain in the Yellowstone National Park at an elevation of between 8200 and 9200 feet (Fig. 4.2), since Antler Peak and Bannock Peak guard the entrance to the cirque.

It is especially because the comb-ridges in the highest levels are precipitous and correspondingly thin that a continuation of the process removes their pinnacles while the broader ridges somewhat farther out and just below the mother-cirques are being sharpened into peaks, both alike through sapping from the cirques.

STAGES IN THE SCULPTURE OF MOUNTAINS BY GLACIERS

To bring together the extremes of mountain glacier erosion which are represented by the Bighorn Range and the Glacier National Park with the intermediate stages which connect them, the four generalised plans of Figure 4.3 have been prepared. In order, these are:

(1) The youthful channelled or grooved upland.

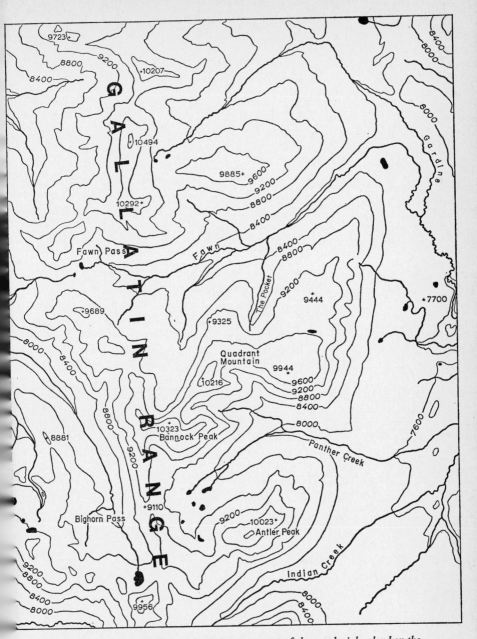

Fig. 4.2   Map of Quadrant Mountain, a remnant of the preglacial upland on the
flanks of the Gallatin Range, Yellowstone National Park
(Altitudes in feet)

(2) The adolescent early fretted upland.
(3) The fretted upland of full maturity.
(4) The monumented upland of old age.

These four stages are perhaps best illustrated by the Bighorn Range, the mass of Snowdon in the Welsh highlands, the Alps, and the Glacier National Park.

The two districts which are here contrasted, the Bighorn Range

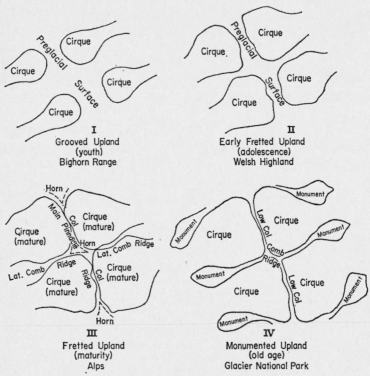

Fig. 4.3 Stages of sculpture by mountain glaciers

and the National Park, also furnish the opportunity to contrast the effects of rock structure in modifying the forms of relief shaped by mountain glaciation. Whereas the high upland of the Bighorn Range has a core of massive rock, thus resembling the Wasatch and Uinta ranges and the Alps, the rocks of the Glacier National Park are sediments and dominantly shales and limestones. It was to be expected that the characteristic structures of these sediments, their bedding planes and their joint system, should exert a strong influence upon the topographic forms produced, as indeed they have. The influence of the bedding planes is displayed in the Glacier National Park in the accentuation of the rock terraces at the upper ends of valleys within the cirques themselves. As in the Canadian Rockies across the international boundary, this character reaches an extreme.

To the well-developed jointing found in the rocks of the Glacier National Park must be ascribed the well-marked checkboard pattern displayed by the park valleys, a pattern which strikes one at once when the topographic map is examined. A number of observations of the bearings of master-joints which were made by the author indicated a rather general correspondence between them and the trends of the valleys in which they were found. In some instances the lower spurs which have been less extensively sapped by glacial erosion indicate very clearly the dominating influence of the joint planes in shaping them. The cirques themselves also display this tendency by their approach to rectangular outlines.

THE TRANSITIONS BETWEEN THE MOUNTAIN GLACIER AND THE ICE-CAP

From the standpoint of the sculpturing of the lithosphere, the ice-cap is sharply set off from all types of mountain glacier through its inability to accomplish a sapping of rock surfaces due to rapid frost-weathering. Its sculpturing processes are therefore restricted to plucking, abrasion, and to a very limited extent frost-weathering on flattish surfaces—processes which in combination leave the rock rounded with surfaces which are flatly convex skyward. That these processes combined play but a subordinate role to frost-weathering in the case of all the types of mountain glaciers, would seem to be sufficiently attested by the sharply accented features which are brought about with their concavities toward the sky.

Since the mountain glacier owes its very existence to a rock container within the lithosphere surface, the enclosing rock walls must in general project above the ice of the glacier. The rock surface will also be reached by air and water wherever crevasses descend through the ice of the glacier to the bed upon which it rests. The conditions essential to the sapping process are a supply of water on the rock surface and oscillations of temperature about the freezing-point. These conditions are not realised either in the case of ice-caps or of continental glaciers, save only where nunataks emerge from beneath the ice near to the glacial margin.

When during an advancing hemicycle of glaciation a mountain glacier is so amply nourished that the rock walls of its containing basins become entirely submerged (ice-cap stage), a profound and immediate transformation takes place in the sculpturing processes. Up to this time, under the dominating influence of the sapping process, the effect of the glacial sculpture has been to sharpen all projecting features of the relief as the glacial basins and channels are carved deeper and extended outward from each individual locus. Now, however, under the plucking and grinding processes alone, which have usurped the functions of the frost-weathering, the pinnacles and horns within the comb-ridges are truncated and ground down, with the result that above the shallowed cirques and the largely obliterated U-valleys there extends a flatly convex surface like that which is fashioned by the same processes beneath a continental glacier. The sharp relief which was inherited from the period of mountain glaciation is thus gradually ironed out into a flatly convex surface ground and polished everywhere by abrasion. The U-valleys are first effaced, beginning at their lower extremities, and the last of the hollowed features of the inherited surface to disappear are the increasingly truncated remnants of the cirques, which in their later stages take the form of an armchair-like depression. Such features are well displayed in Norway where the continental glacier has similarly ironed out the inherited grooved or fretted upland.

Such a surface as succeeds to a fretted upland under the sculpturing action of either an ice-cap or a continental glacier will resemble in form a grooved glacial upland of extreme youth such as is illustrated by the Bighorn or Uinta ranges of the Rocky Mountain region, but it has less pronounced relief and, unlike such a preglacial remnant ('biscuit-cut' surface) the upwardly convex surfaces are here planed and polished by abrasion.

In the receding hemicycle of glaciation which succeeds to the culmination of glacial alimentation, the flat dome of the ice which constitutes the ice-cap will have its surface progressively lowered until the stage is reached at which the rims of the buried cirque remnants begin to emerge from beneath their mantle of ice. In West Antarctica, near the winter quarters of the Swedish Antarctic

Expedition of 1901-3, ice-caps now blanket both James Ross and Snow Hill islands, and, like all Antarctic glaciers, they are in a receding hemicycle of glaciation. On the first-named island the rims of the cirques have emerged from beneath their cover along the eastern and southern margins of the island. The Gourdon and Rabot glaciers are already apparently mainly detached from the dome of the ice-cap, which here rises to its highest point in the Haddington berg. In the largest of the cirques lies the Hobbs glacier, which is still in part fed by two ice cascades situated near the middle of the rim.

Except that the continental glacier, and not an ice-cap, has been the modelling agent, Mount Washington in the White Mountains and Mount Ktaadn in Maine would appear to supply near parallels to the sculpture just described, since the 'gulfs' of the districts have been clearly recognised as cirques. Tarr has claimed that the mountain glaciers which sculptured the cirques on Mount Ktaadn were subsequent to the continental glaciation of the region. This is disputed by Goldthwait, who brings forward evidence to prove that in the White Mountains the mountain glaciers were antecedent to the continental glaciation which shaped the higher and flatter rock surfaces. We hardly see how there could fail to be glacial remnants in occupation of the cirques for at least a brief period while the continental glacier was withdrawing from the region. These would presumably develop in much the same manner as those already described on James Ross island, but with differences which will be pointed below. Goldthwait is no doubt correct in believing that the mountain glaciers had a much longer life during the advancing hemicycle of glaciation and that the cirques were shaped at that time. It is even doubtful if any appreciable work of erosion or deposition was accomplished in the later period of mountain glaciation, and this interpretation would be in harmony with Goldthwait's observations.

THE GLACIAL CYCLE ON THE MARGINS OF THE CONTINENTAL GLACIER OF ANTARCTICA

It is a fundamental and prerequisite condition for the sequence of stages through which mountain glaciers pass during a receding hemicycle of glaciation that the areas of alimentation and ablation should be sharply separated from each other. The former is restricted to the upper levels, and alimentation is augmented in amount toward the top, whereas the area of wastage is found in the lower levels and the losses are increased toward the bottom. Such a distribution results principally from two conditions. In the first place mountain glaciers are nourished by upwardly directed air currents

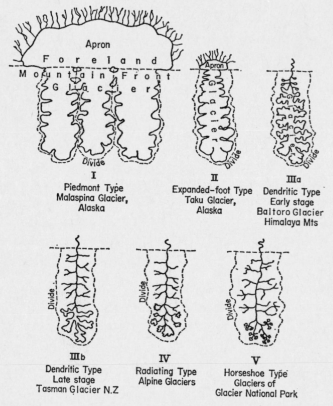

Fig. 4.4   Glacier types of receding hemicycle showing progressive withdrawal of glacier foot

which deposit their moisture as a result of progressive adiabatic refrigeration, and secondly they are wasted by contact with warm-air layers whose temperature rises progressively toward the bottom. It is a direct consequence of the combination of these conditions that mountain glaciers during a receding hemicycle of glaciation become reduced in area through withdrawal of the glacier foot up the valley, and even in its expiring stage the glacier head occupies essentially the same position that it did at the beginning (Fig. 4.4).

Were these two conditions affecting the size of mountain glaciers not realised, the results would be quite different. When we examine the glaciers on the margins of the inland ice of the Antarctic, we find they differ widely from those of moderate latitudes, which are the ones that are well known and have formed the basis of our classification. Within the Antarctic air temperatures do not rise above the freezing-point even in the summer season, save only during short intervals at the termination of the fierce Antarctic blizzards. Furthermore, these marginal glaciers to the inland ice are nourished, not by inwardly and upwardly directed air currents, as are the mountain glaciers of moderate latitudes, but by downwardly and outwardly flowing currents which bring drift snow from the inland ice and often carry it beyond the margin glaciers to be dissipated upon the surface of the sea. Separate areas of nourishment and waste in distribution with reference to altitude are thus not realised, and the otherwise universal law of exclusive drawing in of the foot of the glacier during its waning stages does not hold.

That this is true is particularly well shown in the area of waning glaciers described as 'ice-slabs' by Ferrar, the glacialist of the first Scott expedition to the Antarctic, and fully mapped by Griffith-Taylor, Debenham and Wright of the last Scott expedition. On a far larger scale and related to a continental glacier rather than an ice-cap, these dying glaciers represent a later stage than the marginal types which have already been referred to from West Antarctica—the Gourdon, Hobbs and Rabot glaciers of James Ross island. These glaciers must in an earlier stage have been connected together as a piedmont which was then a part of the parent area of inland ice lying to the westward. From that continental glacier when detachment occurred the rims of the battery of remodelled cirques which rise west of the existing glaciers must have emerged from the ice mantle in forms not unlike those now seen on the margins of James Ross island. Their subsequent diminution in size has gone on through withdrawal both from the cirques and from the lower portions of their valleys—from both extremities toward a central position at a moderate altitude, where the last stand will be made before final extinction.

The usual law of ablation regulated with respect to altitude here plays, therefore, no part, and it is evident that the reflection and consequent intensification of solar heat radiation in the neighbourhood of exposed rock walls has here been the controlling factor in localising the wasting process. This effect of exposed rock surfaces has been recognised for high latitudes by the observation of moats surrounding nunataks and of the lateral streams beside glacier tongues.

REFERENCES

ATWOOD, W. W. (1909). Glaciation of the Uinta and Wasatch Mountains. *Professional Paper, U.S. Geological* Survey, **61**, 1–96

GILBERT, G. K. (1904). Systematic asymmetry of crest-lines in the High Sierras of California. *Journal of Geology*, **12**, 579–588

JOHNSON, D. W. (1899). An unrecognised process in glacial erosion. *Science* N.S., **9**, 106

————— (1904). Maturity in alpine glacial erosion. *Journal of Geology* **12**, 569–587

MATTHES, F. E. (1899–1900). Glacial sculpture of the Bighorn Mountains, Wyoming. *21st Annual Report, U.S. Geological Survey*, 167–90

5 Morphology of Climatic Zones or Morphology of Landscape Belts?

SIEGFRIED PASSARGE

On 22nd and 23rd September 1926 the *Gesellschaft deutscher Naturforscher und Ärzte* met in Dusseldorf. At this conference a series of lectures was given on the morphology of climatic zones. The earth was divided into morphoclimatic zones, and the surface landforms of each area were discussed. While climatic zones are usually delineated on the basis of average values of precipitation, temperatures and humidity, it is to be expected that any attempt to delineate 'morphological climatic zones' will produce divergent interpretations. The obvious question to ask at this stage would seem to be: 'Is climate or are certain climatic zones responsible for distinctive geomorphological features?'

SOME BASIC MISCONCEPTIONS

In my view, two basic errors have been made in the past with respect to the whole problem.

In the first place, the present-day surface landforms are for the most part not the result of present climates, but the product of Pleistocene processes. Consequently the active present-day forces and their morphological expression depend on the climatic zones in which they occur.

With few exceptions climate everywhere has considerably altered since the Ice Age which itself showed wide climatic variations. Now it is an irrefutable fact that natural forces act together and tend toward a balanced relationship. Gradually a condition approaching equilibrium is achieved, so that the effect of any single factor may become significantly modified. But as soon as a change of climatic conditions occurs there is a consequent change in the balance between the various controlling factors. A period of intense erosion, deposition or both may then follow. At present a period of relative calm is being experienced, whilst the protective plant cover, the residual boulder pavement or the gentle slope, for example, severely limit or negate the erosive forces. Thus paleo-features are found, that is the larger landforms, which cannot be explained by forces active today.

Only minor work on river banks or on slopes which have not yet been reduced continues at a modest pace under the present climate. That former glaciation and the period of tundra conditions with its soil flows and solifluction are responsible for many of the present surface features in Germany no one today can seriously deny. The same is true of the 'pluvial' period in the majority of dry areas, but even on the Equator block-debris slopes under the rain-forest, oversized valleys and erosion surfaces covered by swamp prove that the majority of present-day surface landforms are paleo-features. Consequently it would seem appropriate to view the theme thus: 'Surface landforms and their expression in this or that climatic zone'.

In the second place is it, in fact, true to say that present-day morphogenesis is principally dependent on climate?

Theoretically, the answer to this is 'yes', but in practice it is 'no'. Just as in our latitudes the summer is warm and the winter cold, with pressures and winds determining the temperature conditions during these two seasons, so too do the geomorphic forces depend on climate; except that it is the nature of the plant cover, the rock type and even the regolith which in practice determine the effect of the formative processes and the geomorphic development.

An argument against morphoclimatic classification

One example may help to demonstrate why the 'morphological-climatic zones' approach followed at the conference is untenable. The tropics have been divided into three morphological climatic belts: the equatorial with a hot-wet (doldrum) climate; the savanna with a predominant wet season; and the savanna with a predominant dry season. This brief categorisation contains no indication as to how a delineation of the landscapes is to be achieved. The wet-humid doldrum climates are contrasted with the periodically dry climates: the former must, therefore, experience rainfall all the year round. But this 'all-the-year-round' rainfall régime is found only in a few areas of this 'wandering' doldrum belt: almost everywhere a one-to-two month dry season is experienced with at most very little rainfall. Further, in which category do the climates experiencing two wet and two dry seasons belong? Even if we include all climates with a relatively short dry season in our equatorial climate, then we have shown that the suggested climatic classification is unsuitable as a basis for morphological research.

From Conakry to the Ivory Coast the coastal area receives three to five metres of rain and experiences an extreme dry season of four to five months. Without doubt one must regard this area as a 'periodically dry climate with predominant rain'. The plant cover of this area is a particularly luxuriant and dense rain-forest. The soil is weathered to a great depth, and in fact is a soft, moist laterite of cellular structure. Exactly the same erosion processes are present here as in the tropical rain-forests which do not experience a long dry season, that is a vigorous linear downcutting by mountain streams together with landslips, which in conjunction with the formation of bowl-shaped valleys, sharpen the ridges. The features that are formed in this periodically dry savanna with a predominant rainy season are exactly the same as those in the hot-wet equatorial climate. In Upper Guinea, on the other hand, instead of tropical rain-forest, moist savanna with high grass, dry forest and park landscape have developed. Why? *Because here the soil has dried out and slips do not occur in the absence of dense rain-forest.* This hardening of the soil occurs above all in those places where the soil has dried out forming a tough, iron-rich crust in the laterite. Thus, it is the vegetation cover and not the climate that is the determining factor. It is the *landscape* belts and not the *climatic* belts which are important.

LANDSCAPE BELTS

In the montane 'mist forest' (*Nebelwald*) of the tropics, which does not experience a hot-wet climate at all, landslips are even more strongly developed than in the equatorial zone for the very reason that forest-cover and soil-moisture conditions are conducive. All difficulties are easily solved if instead of climatic zones as criteria, one considers the *consequences* of climate, namely vegetation cover, weathered rock and moisture, that is the *landscape belts*. Such morphological landscape belts are readily discernible and comprehensible in contrast to the climatic zones which are seasonally variable and by no means so tangible. While the former are certainly dependent on climate, the morphological differences, as produced by the protecting vegetation cover and the soil type and composition, take priority. Local soils which hinder erosion are, for example, the iron-rich crusts of laterite, lime crusts and moor soils. This example, which proves the inadequacy of the climatically based classification of surface landforms could be repeated many times. One is reminded of some of the basic differences in erosion which exist—those in the tundra with ground ice and soil-creep; in the subpolar zone without ground ice but with powerful earth flow and dissection; and the subpolar forest areas with a long snowy winter, spring meltwater, summer and winter rain with and without ground ice. One only has to consider the effects of ground ice in the formation of extrusive ice surfaces (*aufeisböden*) on floodplains and the effects of ground ice melting on the riverside slopes including collapse of long sections of steep river banks. Further, one might cite the basic differences within the subtropics according to whether one considers sclerophyll forest regions or savannas. Even within this latter zone the presence or non-presence of a limestone crust causes marked morphological differences.

Nebelwald *conditions*

The following example may show the advantages of the landscape morphology approach. According to detailed maps of central and southern Peru, the surface landforms exhibit features due to landslipping. Let us assume that these maps are correct and that they indeed show specific features. Today everything is covered with *borsten* grass and dwarf-bush steppe. The land is partly cold high-steppe with cattle rearing and partly moderate high-steppe with agriculture. There is no ground ice and soil-creep could hardly have developed. The dryness of the climate hardly allows slip to occur, so that it would seem that here we are dealing with large landforms which must have arisen during the Pleistocene period. Now in north and east Peru on the 2500–3500 m-high mountain chains, *Nebelwald* occurs in which, according to Pöppig's work, enormous landslips with steep-sided walls and narrow ridges have developed. Thus these mountains, which are now under steppe, exhibit the features of the *Nebelwald* belt. It appears, therefore, that we are dealing with a paleo-landscape whose essential forms have developed under *Nebelwald* conditions. In Peru there is the moist desert climate of the coast; in the east the hot-wet doldrum climate. Where should one put the high *Nebelwald* climate or the high-steppe? Are we left with anything else apart from *Nebelwald* climate or high-steppe climate? Would not this be merely a superfluous description? Surely it is easier to speak of 'the surface features or surface configuration of the *Nebelwald* or the high-steppe', that is to use landscape terminology. Moreover expressions such as 'surface features of the interior and high deserts' imply landscape concepts. Why do we go only half way?

Landscape morphology will inevitably take the place of the morphology of climatic zones, because it succeeds in describing and explaining the phenomena far more simply and clearly. Obviously not all difficulties arising from transition zones, local soils and local vegetation will be overcome as soon as landscape morphology is introduced, but in any case the result will be far more satisfactory than if one bases morphological research on climatic zones.

6 Landforms of the Savanna Zone with a Short Dry Season

FRANZ THORBECKE

When I speak of the landforms of the 'wet' savanna, I refer to tropical areas in which wet and dry seasons are clearly discernible, and in which, therefore, the dry season lasts for such a time that the characteristic features of a dry climate are so clearly developed that they become morphologically important. In the dry season rain must entirely cease and the temperatures must be so high that the ground really dries out, high enough, in fact, to cause mechanical weathering of the rocks. The air, too, must be dry. This immediately excludes areas like western Guinea, the Ghat area of India, and the Atlantic margins of Brazil, in which the monsoonal air movement coincides with a steeply rising coastline or a humidity which penetrates a long way inland, so that forests similar to the equatorial zone occur.

A transition zone between the continuously wet area and the area with which we are dealing here might be where a dry season of two to three months occurs. This, however, is not morphologically effective because, during the rainy season, the ground is thoroughly saturated and also because rain still falls occasionally during the dry season. This means that the ground, at least in the lower levels, remains somewhat moist and the air relatively humid. I have observed that the forests which grow under such conditions are distinguishable from the equatorial rain-forest by the fact that they have little or no undergrowth. It is important to study more closely the morphological processes in the transition zones, for as far as I know there are no observations on the subject.

I know of hardly any observations from America or Asia on the relationship between surface landforms and a climate which is characterised by a predominantly rainy season with, however, a morphologically effective dry season. In India we know of the same characteristic landforms that I have observed in the corresponding climatic zone in Africa, but as far as I know English and Dutch geologists have not examined the climatic effect in causing landforms. I therefore base my paper on my own observations in the periodically wet highlands of the middle Cameroons.

This interior highland region, a mirror of the Brazilian highland, is a high grass savanna with isolated trees and forests which follow the river banks. The climate is sharply divided into wet and dry seasons. In the middle of the nine-months wet season a slight drop in precipitation is observed which, however, has no morphological effect. The dry season lasts generally about three months, but even in this short time exhibits a marked dryness which is very important in terms of morphological processes.

THE MORPHOLOGICAL EFFECTS OF CLIMATE

Mechanical weathering

The inner highlands of the Cameroons are composed mainly of granite and syenite, with some gneisses. To the west lie basalt plains, above which igneous massifs tower.

The granites and syenites are especially susceptible to intense insolation and a wide temperature range. Their layered form facilitates the action of atmospheric forces: the layers expand during the day and contract at night. The total effect is a shattering of the layers; this can be of centimetres or even a metre thickness depending on the original mass of granite. I observed that on steep rock faces this shattering occurred firstly at the base and then moved upwards, the sharp edge of the upper part of the layer remaining fixed to the rock wall. It does not fall, but is gradually washed and rounded by the rain in the following rainy season, until the rock face again appears smooth, having receded by about the thickness of the layer. I observed each stage of this process at all heights on the rock face.

The material that falls collects at the base of the face and decays there, by the extensive and extremely quick effect of insolation, breaking down into the smallest pieces. I have never found blocks or boulders at the foot of individual mountains nor on steep rock faces. True, they do occur near mountain peaks, where shattering is particularly rapid due to increased insolation and cooling. But at lower levels large and small blocks are so rare that the locals carefully guard their hearth stones, mill stones and anvil stones, because it would take a long time to find a replacement. Debris of the size of a fist up to head size, so characteristic of our latitudes, is completely lacking here.

As a particularly striking example of a single occurrence I observed the formation of solution grooves (*karren*) on a syenite wall on the highest rock peak of an inselberg massif. The *karren* were hollowed out to a depth of several centimetres and covered the whole wall. I suspect that the particular nature of the rock, in association with the torrential rain, caused this minor feature, which is completely alien to the area.

On the horizontal surfaces of large rock layers, both in the plains and in the mountains, a number of flat, panlike depressions occur: they are often circular and rarely occur singly. I myself have seen one with a diameter of a metre and a depth of about 15 to 20 cm. Smaller *pfannen* (pans) measured 20 to 30 cm across and were quite flat. Often the pans and areas between them were covered with lichen. Once, in the bottom of several small pans, I found the coarse felspars of the syenite which were the remnants of the thin layers which had been shattered. I feel that a mixture of mechanical weathering during the dry season, together with the 'exploding' effect of the lichen which grows in the cracks of hairline thickness and the rainwater which collects in the minute depressions is responsible for this occurrence.

Chemical weathering

Chemical destruction during the long rainy season is increased by the high saltpetre content of the thunderstorm rain. The debris is quickly altered into lateritic material.

Even level or moderately sloping rock surfaces are attacked in places by moisture down to several metres and so changed into laterites. The end result of such short but intensive mechanical weathering and lengthy and powerful chemical destruction is the complete lack of any coarse-grained weathering products. The surface consists either of a rock that has not been attacked, or one completely replaced by laterite.

The work of sheetwash

The second effect of the rainy season is the transportation power of water in mighty sheetfloods. They tear away the small weathered particles from every slope and carry them long distances. The storm channels of the first rains have the greatest effect. I observed how fine white sand (the unaltered product of mechanical weathering) which lay at the scarp base was carried out far into the plains and there deposited as a shimmering cover over the red loam of the weathered regolith. A village square in the same area was covered to a depth of several centimetres after one such powerful flood. The weathered material is also stripped away in places by surface wash. On every slope the 'stumps' of savanna grass are nearly a quarter of a metre high, indeed in some places up to half a metre. The slopes which have a covering of weathered material may also be preserved to a fair extent. Despite the mesh of roots, however, the weathered ground is often so quickly eroded away that parts of the rock wall even at the foot of the slope are continuously being laid bare. On both the isolated mountains and the plateau scarps I found at all heights of the slope smooth, bare rock walls many metres high, surrounded above and below, to left and right by grass-covered soil. As I never found any altered material that had fallen recently I do not think that large slips take place, but more that one fairly lengthy process is responsible.

Neither have I ever observed soil-creep on the grassy slopes. The relatively dense savanna grassland helps to hinder the sheetfloods, but the tremendous amount of water and the duration of the rainy season more or less completely balances this: thus the morphological effect of sheetflood in the savanna does not compare with that on the more thinly covered steppe.

Where the ground is moist, as a consequence of increased rainfall or a nearby stream, and there is tree growth, similar weathering and erosion forms occur as in a climate that is continually wet; only in these places, especially in stream gullies, can soil slip and slides be observed. In the plains the parent rock often alternates with deeply weathered soil profiles, for the most part red laterite. Gigantic granite blocks and large piles of weathered material lie right next to each other at exactly the same height; a rock hollow is filled with red and yellow clay, formed mainly from *in situ* weathered material, but some is also carried down by the sheetfloods.

The work of running water

The great amount of running water is clearly shown by the large number of water channels which exist; large and small rivers are fed by innumerable streams which descend from all heights. The river density is so great that many slopes appear as one great continual water course. In the rainy season all the gullies and rills between the permanent streams are filled with rushing water. All this running water joins up at the foot of the hills with the larger rivers which, for long stretches, run parallel to the base of the mountains. The combined water power erodes away the ground at the base of the mountains and plateau scarps; streams flow from the mountains down the plains which are steepest at the mountain foot but distally become more gentle. More quietly flowing water gives rise to swampy conditions and these are distinguished from the rest of the countryside by their light green colour. The plateau areas, too, are covered with water channels, even the smallest channel being cut deeply into the surface. Springs rise on this surface and are incised some four to six metres. A steep clay wall surrounds a spring of about five-metre diameter. At its foot appear streams which although only the merest trickle during the dry season never completely disappear. The gorges of the streams are filled with dense forest, which forms along the banks and towers up above the level of the gorge. In some places there are steep-sided treeless channels of about one metre in width and incised two to three metres: these are dry in the dry season, yet contain water during the rainy season. The water which rises at the spring usually has a winding underground course above the spring point. The clay covering above gradually sinks along the underground watercourse.

Since the sides of the gorges are always so steep, there is no marked valley asymmetry. Not only near the springs but also over large areas the gorges retain this steep-wall characteristic—proof of the fact that the side walls always collapse steeply and so quickly that surface wash does not transport the debris and so modify the slopes. Because of the continual competitive erosion by streams, the interfluves on the plateau surfaces are constantly being changed and lowered.

THE LANDFORMS

The interaction of all the processes creates characteristic geomorphological features. In youthful areas, for example, on the slopes of volcanic plateaus and mountains and on the edge of the inner Cameroon highlands, these forms are not so obvious. The large amount of surface water leads to a dense valley network and the large amount of runoff causes rapid vertical and lateral erosion. The effect of sheetwash on the slopes is also evident, the slopes remaining generally relatively steep, whilst a broad valley meadow is quickly formed, lacking completely, however, in *schotter*. With the cutting of such steep valleys, plateaus must be formed in which steep mountains remain, but I have never observed this.

The full morphological effect of these processes is not felt, how-ever, until the inner part of the highlands is reached and the in-fluence of the 1000-metre high edge effectively ceases. Vertical erosion becomes less as lateral erosion increases. Rock faces are not covered with scree but remain steep. Mountain slopes consisting of weathered clay retain their steep slope and are not lowered. The mountain edge rises abruptly from the surface of the plain, which continually increases its extent and eats away at the residuals. Even the mountains lose area to the plains and their rivers.

The stepped landscapes

As a result, extensive erosion surfaces appear as plains whose hol-lows have been infilled by weathered lateritic material. Individual mountains and scarps at the edge of higher surfaces, which them-selves demonstrate the same features, rise above this erosion surface. Along and beyond the scarps separating one surface from the next one below it there occur bare rock mountains which show clearly that they were previously connected with the upper surface. They are buttresses occupying a former position of the scarp. Deep gorges are cut into this scarp and the back wall indicates the future course of scarp retreat. The present outer bastions are only linked to the upper surface by narrow rock strips. A row of outliers will be formed in a geologically short time. The initial formation of such a rock step is in this context unimportant. Whether it is due to faulting or folding or a result of the forces of erosion, the present edge had been eroded back by presently active forces and will continue to be attacked in such a way by these forces and eroded further back. Thus features appear in the hard crystalline rocks which are similar to our scarp-lands but are the result not of the variable resistance of stratified rocks but of a seasonal climate.

The inselbergs

If, in an area in which vertical erosion no longer predominates, the higher areas are simultaneously attacked on all sides by powerful river systems and continually exposed to slope weathering then from a stepped landscape appear individual, sharply rising massifs called *inselbergs*.

I was fortunate enough to observe the actual formation of a middle-sized inselberg, which lay in a line with other inselbergs. A gentle west-to-east sloping surface is attacked at its western edge, the weather side, by erosion forces which cut deeply due to a base-level some 100 m lower down. The steep edge is gradually eroded into a series of separate steep mountains, in between which stretch extensive flat areas. One such inselberg, however, is still joined to the main area. Gradually climbing the eastern side of the sloping surface

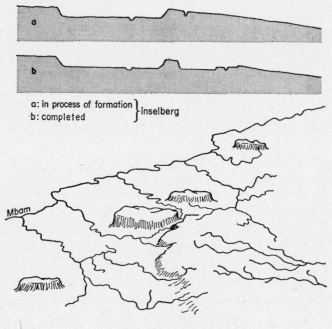

a: in process of formation ⎫
b: completed   ⎬ inselberg

Mbam

Fig. 6.1   Inselberg formation

I suddenly found myself standing at the same height above sea-level as the inselberg without having realised it. A narrow gorge, 100 metres deep with steep sides, separated me from this outer bastion which was in the first stage of separation. Looking back later the surface of the inselberg and the erosion surface which sloped gently towards the east appeared as one plateau, showing how little effect the separation of the bastion had on the landscape as a whole. Two streams cutting deeply in from south-west to north-east had brought about this separation.

In a climate such as our own, this kind of separation would take place after a broad, gentle lowering of the intervening neck of land. In a climate with a long rainy season and a short excessively hot, dry season, however, the initial stage of deep incision is immediately followed by a lateral retreat of the steep walls. The same effect was noticed on the plateau between the steep walls of the inselberg, which had just been formed, and the walls of the inselbergs which were obviously about to be formed. Further downcutting by the streams and the lateral extension of the gorge into a plain will isolate the inselberg in a relatively short time. It will be formed literally before our eyes and, at the same time, its present extensive surface, which provides space for three villages, for agricultural land and forest, will similarly be diminished as those of the neighbouring, clearly defined inselbergs have been. If this decrease in area continues, then eventually only inselbergs of the smallest size will remain, looking more like huge blocks than mountains but which, nevertheless, possess all the morphological characteristics of our inselbergs.

Here then, in a periodically dry climate with a long rainy season and a short, but excessively dry season, features both small and large and differing from each other, in gradational fashion, are formed by the forces of the present-day climate.

7 The Climatic Geomorphic System

JULIUS BÜDEL

EVERY landform on the earth's surface is dependent on three major causal factors—endogenic processes, exogenic processes and the local geology. The endogenic processes form the large basins and ranges of the earth's surface, that is the basic features which deviate from the true spheroid earth surface by no more than 10 km in the vertical plane.

Individual landforms of the earth's surface, however, are the result of exogenic processes, caused by climatic conditions and ultimately by solar energy. It is these processes which transform the endogenic features, by weathering and erosion in the highland areas and deposition in the basins, into the extremely varied collection of landscape features which characterise the earth's surface. From the beginning, therefore, it must be supposed that landform variety closely follows the major climatic zones. However, before embarking on the general application of this relationship, one special case, that of mountains, should be noted. Mountain areas exhibit an altitudinal zonation of climate: the forest zone and snowline are to be found at completely different heights and in each zone the individual landform elements take on climatically controlled characteristics. Each mountain area preserves its own character, therefore, even if this is controlled by complex climatic conditions. From this we can clearly regard the mountain zone as exceptional, and, as we are here concerned with the morphological influences of the major climatic zones, we may exclude mountain regions from further consideration.

If one were to confine oneself to the study of a small area, as is usually the case in geomorphological research, the various geological factors would appear dominant because climatically caused differences would not be detectable in the landscape. However, when larger areas are compared, the importance of the geological factor declines greatly. If, moreover, observation is limited to areas of flat and gently sloping relief (by far the majority of the earth's surface), then we see that geologically caused differences in landscape features are far outweighed in importance by those due to climate. Accordingly, the latter appear more appropriate, as a systematic foundation for geomorphological study.

The aim of the following observations is to gather together into a system all these climatic influences on landscape. It is an attempt to present an overall view of a problem with which I have been involved for the past 18 years (Büdel, 1944, 1948, 1949). Of course, any such first attempt must contain deficiencies. The system seeks neither to minimise the acknowledged relationship between landform and geological structure nor to discredit the methods and points of view which have been responsible for the extraordinary success of classical 'genetic' geomorphology. It merely wishes to place the greatly neglected climatic factor against the geological, so as to establish a better balance in geomorphological studies.

PENCK'S CLIMATIC SYSTEM

Of course, the climatic factor, in some form or other, has nearly always been incorporated into geomorphological systems. The most famous is the climatic system which Albrecht Penck (1910) introduced into geomorphology and which, since Davis's contributions, has become a major pillar of the subject. Basically, this has not been replaced by any other system. According to this system, the earth has three main climatic zones: the nival (the effect of glaciers), the humid (river action) and the arid (neither glacier nor permanent river action, but powerful wind action) and each of these zones displays a climatically influenced landscape.

The importance of the Penckian system lay, above all, in the fact that it appeared to exclude completely the possibility of climate influencing landform variety in any detailed way. This is simply not realistic, of course. At least six large areas within the humid zone can be distinguished climatically, and I have already suggested that such areas might be termed climatic–morphological zones (Büdel, 1944). Their individuality deserves just as much recognition as the arid and glacial–nival features. With these and the submarine landforms we can now divide the earth's surface into at least nine large climatic–morphological zones instead of only three, and these in turn can be subdivided. Clearly, there is no comparison with the Penckian system, even if one subdivides the proposed major divisions into semi-humid, semi-arid or semi-subnival. The reason for this is that the Penckian system lacks a morphological basis. Fundamentally it describes something completely different—namely the main types of terrestrial water régime. It does not deal, as the title of Penck's work states, with a climatic division on a physiographical, or even on an all-round climatic basis, but rather on a specialised hydrographical basis. Accordingly, the system is comprehensive only in the case of the nival climate, in so much as glaciers attack all the land they cover. In all other parts of the earth, however, Penck's subdivision establishes only whether or not rivers are present, that is one linear element of erosion. It is exactly such distinctions which are responsible for the occurrence within the humid zone of a very varied array of climatic landforms. Moreover, they affect the landscape as a whole and not merely the erosion channels.

Surface denudation

Climate is responsible for the type of surface denudation activity as well as for the erosional processes in a given area. These are related to climate in a rather different way than are the Penckian types of water régime. The surface denudation processes are dependent on the type of soil cover. In many parts of the world these are more dependent on climate than on parent rock and, indeed, the result of a very varied and complex climatic interaction. We are concerned here, however, less with the climatic factors of soil formation than with the collective sum of climatic interactions which set the soil into motion: whether this motion is powerful or weak, and whether it affects just the uppermost soil particles or is present at depth. The distribution of various denudation types must be linked even closer to certain climatic zones than the soil types, and obviously to a greater variety of types than the basic nival, humid and arid climates.

A climatic classification which accounts for only a few geomorphic processes and which provides no classification of denudation features is unlikely to lead to a comprehensive division of the earth into climatic–morphological zones. As Mortensen (1930) stressed, such a classification cannot be made deductively but only inductively, that is the classification must emerge from a consideration of the morphological features themselves. Only when we have successfully delineated certain areas with similar morphological characteristics, can we begin to investigate the climatic characteristics which give rise to this similarity.

CURRENTLY ACTIVE LANDFORMS

This method is far more complex than might at first appear. One cannot simply base one's findings on the larger, more obvious landforms. The reason for this is that in most parts of the world the present landforms are not the products of present-day processes, and therefore cannot be related to present-day climatic conditions. The majority of surface landforms, especially those caused by the lengthy processes of denudation, are extremely resistant and alter far more slowly than the climate. Geological history has witnessed a succession of climatic changes. For example, since the early Tertiary period climate has gone through all the zonal varieties now found from equator to pole. It follows that if we are seeking to establish definite relationships between present-day climate and landforms, then we must consider not the large landforms which are a remnant of numerous climates, but rather the currently active processes and forms. If this is done, then further investigations are necessary to show which characteristics of the landforms are caused by present-day morphogenetic processes and which (in a more or less obvious form) still show the inherited characteristics of former climatic–morphological circumstances. The most important problem posed, however, is not the separate recognition of the predominant processes of surface and linear erosion, but how these two groups of processes interact. It is precisely in this way that the individual climatic landform areas are most clearly characterised (Büdel, 1948). Perhaps the most obvious correlation is with certain climatic factors. However this is not to say such relationships can be established with individual instrumentally recorded meteorological elements. The establishment of a relationship between selected meteorological elements and certain geomorphic processes is much more difficult than it may at first seem. To find a simple mathematical correlation is hardly possible at present. But even a simple demonstration of spatial correlation between such features is not easy. Only rarely will the boundary of a particular process coincide with such values, in the way that Penck (1910) or Köppen (1936) have used to delineate their climatic zones. Furthermore it is the quite special, apparently trivial characteristics of the climate, which often determine the distribution of a geomorphic process; characteristics, which even in the developed countries either cannot be measured accurately enough or cannot be measured at all. Thus, for example, according to Troll (1944) the occurrence of solifluction due to daily freeze-thaw in middle-latitude mountains depends on how many of the freeze-thaw cycles occur in the snow free period. In sum, the main difficulty appears to lie in the establishment of causal and spatial relationships between climatic and morphological features.

More promising is the correlation with complex climatic effects, which influence geomorphic processes and which in turn are themselves influenced. One is left with the whole complex of processes of climate, morphogenesis, soil formation and plant cover. As long as such a complex remains stable then we may expect to recognise a particular climatically determined landform area. If this relationship

Table 7.1   Climatic–morphological zones of the Earth

Submarine morphological province = sea floor zone
Subglacial morphological province = glacier zone
Subaerial morphological province, comprising:

1. Frost debris zone
2. Tundra zone
3. Temperate mature soil (*ortsboden*) zone

   (a) Oceanic zone
   (b) Subpolar zone, without *tjäle* (permafrost)
   (c) *Tjäle*, or permafrost, zone
   (d) Continental zone
   (e) Steppe zone

4. Mediterranean transition zone
5. Dry debris zone

   (a) Tropical hot deserts and marginal deserts
   (b) Extratropical deserts
   (c) High altitude (cool) deserts

6. Sheetwash zone

   (a) Tropical
   (b) Subtropical

7. The inner tropical zone of mature soils

---

alters, then a new one begins to develop. I have used the term 'climatic–morphological zones' to denote the complex nature of all the processes that are vital to morphogenesis in a given area.

Since soil cover and plant cover play an important part in this complex interaction and since both these elements can be greatly influenced by man, then man can influence the balance between them and so exert a noticeable influence on the morphological processes. Thus, in contrast to the paleogeographical features and the 'natural' condition of the geological present, humanly induced processes must be considered in any general consideration of climatically determined landform zones.

On the basis of such principles we can now divide the earth's surface into nine major climatic–morphological zones, themselves subdivisions of the earth's three major natural provinces: the submarine, the subglacial and the subaerial. We shall not subdivide the submarine province in which practically all climatic influences are lacking, nor the subglacial province, since here there is only one climatic influence, namely the process of glacial erosion. Our main sphere of interest lies in the subaerial province, and this is subdivided into seven major climatic–morphological zones, the majority of which are further subdivided (Table 7.1). Clearly, for some zones, in particular the tropical and subtropical, there is a dearth of exact observations. It is perfectly possible that in time, one or other of these might be considered a major zone in its own right.

THE SUBGLACIAL ZONE

The subglacial zone is the only one that coincides fully with one of Penck's climate areas, namely the nival. It is exceptional, because only one formation process—glacial erosion—is in action and this involves both surface and linear erosion. In fact we do distinguish between glacial denudation and glacial erosion, depending on whether the ice is flat (plateau glacier, ice sheet) or linear (valley glacier) in form, but here both expressions are used as stages of one and the same process. Its quantitative capacity (ability to transport material in a given time) depends on the thickness and speed of the ice flow (Louis, 1938) and in individual cases this can become extremely powerful. The resulting landforms are qualitatively unique and such important and long recognised features cannot possibly be confused with the morphological features of subaerial erosion in other zones.

THE SUBAERIAL MORPHOLOGICAL PROVINCE

The frost debris zone

The frost debris zone and the tundra zone were formerly grouped together as one under the title of the soil flow zone (Büdel, 1948), but it now seems better to separate them. Both are characterised by solifluction, a powerful denudation process peculiar to this zone.

The soil flow area is sharply demarcated on the poleward side by the glacier zone, and equatorwards by the circumpolar forest zone. But their subdivision into two zones is still quite obvious. The frost debris zone, which is nearer the pole, is the area of vegetationless, cold, rock desert. It is characterised by unlimited soil movement, with striped ground on all slopes between 2° and 15°. On steeper slopes, however, it gives way to two other even more important erosion processes, namely gully wash, predominant on slopes between 15° and 40°, and slope shattering on gradients of 40° and above. Because of the presence of these processes, not only is much coarse debris carried down to the streams, but the runoff water reaches them efficiently. The bulk of the annual runoff occurs during the short snow-melt period, when evaporation and percolation losses are negligible. Therefore, streams are capable of powerful erosion both of the lateral and the vertical kind, that is, all streams in this zone not only have broad, debris-covered floors but they also succeed in deepening their channels. These debris-covered channel beds characterise the river to some extent right into its highest reaches where they usually become steep-sided and gorge-like. Small basin-shaped valleys, evidence of rather weak linear erosion, are almost completely lacking in these higher reaches. Thus this zone is characterised by a unique combination of powerful erosion and denudation processes, which can operate completely unhindered by plant cover and which can mutually strengthen each other. In quantitative terms, the erosion capacity per unit of time is greater in this zone than in any other subaerial zone: it even exceeds that of the glacier zone (with the exception of a few special cases). In qualitative terms, the frost debris zone occupies a special position due to the uniqueness of its features, although it is perhaps less distinctive than the glacier zone and the sheetwash zone.

The tundra zone

The geomorphic processes of the tundra zone are similar to, but generally weaker, than those of the frost debris zone. One important reason for this is the relatively thick plant cover. In vegetation-free areas, solifluction results in irregular solifluction terraces. Although weaker, however, this process extends over a greater range of slope than in the frost debris zone, namely on all slopes from 2° to 20° and occasionally even on slopes up to 30°. Thus the exposed surfaces in the tundra zone are severely restricted, in contrast to the frost debris zone. Moreover, the solifluction terraces have a tendency to block the runoff, so that not only less debris, but also a lower percentage of precipitation reaches the rivers. It follows that stream erosion potential is less, the debris-strewn channels are becoming narrower to such an extent that, in the upper reaches, they are often replaced by small corrasion valleys, or *dellen*. In this zone, currently active major landforms develop much more slowly than in the frost debris zone. Nevertheless erosion capacity in the tundra zone remains considerable and exceeds by far that of our moderate central European climatic zone.

The temperate mature soil (Ortsboden) *zone*

Under present conditions this zone is larger than the two previously mentioned zones and includes the extensive mid-latitude forest belts. Also included in this zone are the wetter, densely forested peripheral parts of the extratropical steppes, the transition to the extratropical desert-steppes and deserts occurring very gradually. This zone, therefore, incorporates the belt of podsolised grey and brown forest soils and also the so-called black-earth zone, whilst the area of the chestnut and grey steppe soils must be regarded primarily as a climatic-morphological transitional area between the temperate zone of mature soils and the nontropical dry debris zone (*Trocken-shuttzone*).

In sharp contrast to the frost-debris and tundra zones, fairly thick soils cover the bedrock from which they have developed *in situ*. On all slopes between 17° and 27° (the angle varying with the type of rock), the effect of the denudation processes on these soils is confined to a thin surface layer, often only a few millimetres thick. The removal of particles is not so much by mechanical erosion as the result of solution (often wrongly called colloidal) and of chemically released substances which pass into the runoff water. As a result of this limited action not only fossil landforms, but even fossil (Quaternary) weathering layers are found over the whole of this slope range; these have undergone slight chemical change, but no mechanical transport of any kind. Where the natural vegetation cover has remained undisturbed, only very rarely has the soil cover as a whole moved, and then only where soft rocks have been affected by excessive percolation. As slopes increase from 20° to 30° up to 50° and 60°, the more powerful mechanical processes gradually gain the upper hand, although, in general, they do not match the capacity of the weakest denudation process of the tundra zone, namely solifluction. That only occurs when we cross the treeline and enter the moss, lichen and *felsenmeere* areas of the mountains.

Even the erosive capacity of the rivers is fairly slight in this zone. The weak denudation processes deliver little coarse debris to the streams (except in the higher mountains), the difference between high and low water stages is small because of the equable climate, and runoff as a whole is reduced by the dense soil and vegetation cover which takes up a large percentage of water by evaporation and percolation, thus making it morphologically ineffective. Even the higher floods which occasionally occur leave only a thin covering of mud on the floodplains: in postglacial times this has gradually developed into the well-known alluvium layer (*Auelehmdecke*). Only rarely in postglacial times, and then only in the case of large rivers with steep gradients, has greater deposition occurred on the floors of the Quaternary valleys (Büdel, 1944; Mensching, 1950). In contrast to general surface denudation, even such relatively weak erosive action appears to have noticeable capacity, and most of the important postglacial changes in the landscape can be attributed to it. More important than changes in the few big rivers is the fact that linear erosion in quite small streams has everywhere notched V-shaped valleys (*Kerben*) into the broad Quaternary valley floors.

In such an equable climate, the temperate mature soil zone is quantitatively the least dynamic and qualitatively the least distinctive of all these zones. Since erosion capacity is small more time is required for it to impress its specific morphological character on the older, inherited paleo-forms. Even given man's powerful interference with the climatic-morphological balance, the erosion processes have been accelerated in only a few places.

The conditions described are primarily valid for the central part of this zone, that is the central and west-European and north-American deciduous and mixed forest zone. In the boreal conifer zone of central Sweden and southern Finland, the effects of erosion slowly begin to increase northwards, especially as a result of powerful floods fed by annual meltwaters. On the other hand, the significant strengthening of the denudation processes as shown by recent solifluction features becomes evident only very close to the polar forest limits (the subpolar zone without permafrost of Fig. 7.1 and Table 7.1, approximately delineated by the subpolar pine and bog forest). In comparison with the tundra zone, then, the temperate zone of mature soils is quite sharply defined in extent. This border becomes less sharply defined further east, and beyond the Yenisei River the area of permanently frozen ground suddenly runs southwards deep into the forest belt. Here, within the middle-latitude forest zone, a greater tendency towards deep-seated soil movements due to frost is again evident. Little systematic work has been done on the importance of this process for the type and extent of surface erosion. These areas of forest with permanently frozen ground may be termed the zone of mature soils with *tjäle*.

Even south of the area of perennially frozen ground, however, a slight change of character occurs on the continental side of the zone. This is caused primarily by the increasing effects of seasonal flooding. In the permafrost-free forest zones of eastern Europe, Siberia and the corresponding belts of North America, one can speak of a 'continental temperate mature soil zone'. The pine forest region of southern Fennoscandia must be included in this group (Fig. 7.1). The types of denudation are approximately the same as those in Central Europe. Crossing into the extratropical steppes and prairies, however, the surface processes alter markedly and, indeed, change to a progressively greater extent the further one goes from the woodland steppe (*Waldsteppe*) across the feather grass steppe, into the desert steppe and finally into the cold inland desert itself. The steppe area is a large zone of transition in which there is progressive increase in the efficacy of the geomorphic processes from the margins of the temperate mature soil zone towards the extratropical desert debris zone. A sharp climatic–morphological boundary, like that of the polar forest zone, is lacking. The major surface characteristics arising from these increased erosional processes are the steppe gorges (*Steppenschluchten*), for example, Balkas and Owragi (Schmidt, 1948). Although individually they are small erosion features, we must include them in the array of surface forms because of their widespread occurrence, like the slope gullies (*Hangrunsen*) of the frost debris zone. (At the same time, next to the more steeply incised, certainly more recent of these notches, numerous gently hollowed landforms occur; these are extremely reminiscent of the small Quaternary corrasion valleys called *dellen* in Central Europe and which are presumably of Quaternary origin, although up to now it has not been possible to separate with certainty fossil and recent features of this type.) If, under present conditions, the tendency towards the formation of such gorges is decreasing, as seems to be

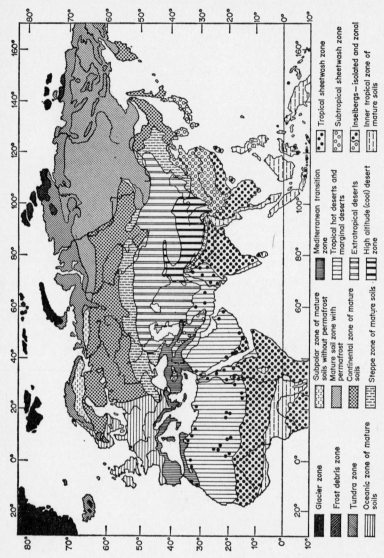

Fig. 7.1 *The climatic–morphological zones of the Old World. The exceptional climates of the high mountain areas .... are into consideration only polewards of 60°N*

Glacier zone

Frost debris zone

Tundra zone

Oceanic zone of mature soils

Subpolar zone of mature soils without permafrost

Mature soil zone with permafrost

Continental zone of mature soils

Steppe zone of mature soils

Mediterranean transition zone

Tropical hot deserts and marginal deserts

Extratropical deserts

High altitude (cool) desert zone

Tropical sheetwash zone

Subtropical sheetwash zone

Inselbergs – isolated and zonal

Inner tropical zone of mature soils

suggested by their number and distribution, then such instability clearly indicates aberrations in the climatic–morphological balance. If, in spite of this, we include the wetter parts of the dry zone (forest steppe and part of the feather grass steppe) in the temperate mature soil zone and call it temperate steppe then it is for the following reasons.

(1) *Orts soils* (those characterised by some leaching or by accumulation of salts or organic material in the profile) extend throughout the whole steppe region, even as far as the edge of the dry debris zone.

(2) There is a marked difference between the wetter and drier parts of the steppe zone, especially in respect of man's interference with the climatic–morphological balance. In the wetter steppe zone such interference leads in general only to single features (important road works, the occasional ploughed field for instance). It is only in the drier steppe zone that such influences may initiate extensive, self-perpetrating soil destruction, whether soil erosion (the formation of badlands) or surface deflation.

(3) In the wetter steppe zone, present-day erosion processes, despite their reinforcement by man, are not significant outside the Central Eurasian region. This can be seen from the fact that the vast loess covering of the Ukrainian steppe has not experienced a particularly more severe or markedly different type of erosion than the Central European loess. It was this consideration especially which made me include the densely forested borderlands of the extra-tropical steppe (roughly the area of the black earths) in the temperate zone of mature soils.

The Mediterranean transition zone

If the extratropical steppe, as the transition zone between the temperate mature soil zone and the interior deserts, exhibits a marked strengthening of the geomorphic processes, then the Mediterranean type of climate, as a transition zone between temperate mature soil areas and coastal deserts displays a still greater qualitative change and quantitative increase in these processes; this seems to justify the recognition of it as a separate climatic–morphological area. Its individuality increases with proximity to the desert.

From the purely climatic point of view the Mediterranean zone has greater similarity to the desert than to the extratropical steppe, despite its original, extensive forest covering. Frost soils, snow and

ice are not experienced so that snow meltwaters are absent. Winter
are mild and rainy and plant growth is hardly interrupted, a situation
which accounts for the original vegetation cover of forest. In summer
a really desert-like dry climate is experienced, and the summer i
longer the further south one goes. The dry summer and the sharp
rainy season are the most significant characteristics of morphological
importance in this climatic area, and they explain its individuality.

As in the extratropical mature soil zone, denudation on gentl
slopes is predominantly by chemical solution and the transportation
of particles in suspension by percolating rainwater. But since the
Mediterranean forest types are lighter and have a little less ground
vegetation than those of Central Europe, there is also an increase in
the mechanical action of surface wash. Even in flat areas this may be
sufficiently intense to result in the formation of steep-sided rills and
gullies (*Kerbspülung*) which resemble the steppe gorges of the Ukraine
On steeper slopes, the gullies become larger and more numerous
In mountain areas they can even form stream systems, which then
proceed to dissect the slopes. In certain cases this phenomenon can
reach such an intensity (especially in the southern Mediterranean
that it resembles the conditions found in the frost debris zone and in
the neighbouring areas of the dry deserts. In clay areas landslipping
and debris flow are more common: these represent a well-known
erosion complex. In contrast to the soil-mantled karsts of lower and
medium altitude areas of Central Europe, the Mediterranean kars
displays very many well-defined limestone features—*karren, dolines
uvalas, poljes*.

The uniqueness of the Mediterranean zone is demonstrated no
only by the surface processes, but to an even greater extent by the
linear processes. As we have seen in the temperate zone of mature
soils (including the nontropical steppe), morphologically significant
river action occurs mainly in the central channel, which carries flow
throughout the year. This channel has usually cut a few metres at the
most into the broad valley bottom created during the last glaciation
These recently established river courses appear to change position
very little (apart from the largest rivers and rivers with considerable
gradients). Even at high-water stages, their erosive activity is largely
restricted to the main channels, while only a thin layer of mud is
deposited on the valley floodplain. However, the general form of
river valleys in the Mediterranean zone would be considered exceptional if found in Central Europe. Mediterranean river valleys are

similar in form to those of the soil-flow zone. The broad alluvial floodplains are absent. Furthermore, the rivers do not receive constant rainfall and therefore do not have a principal, perennial water channel. In winter, the season of highest-water stage, sparse plant cover enhances the rate of runoff, the rills and gullies evacuating the ample rainwater directly to the rivers. They exhibit a broad flood-stage bed, which fills the whole valley bottom. This is utilised to its full extent by the river, with braiding and the deposition of gravel bars usually taking place. However, as debris production is far less than in the soil-flow zone, the flood debris in the valley is neither as extensive nor as sandy as in the latter zone, although it may be deposited more thickly over the whole breadth of the channel (Mortensen, 1927). The debris-strewn torrent bed with its steep, undercut banks is a special characteristic of all Mediterranean-type areas. The bed is dry and shining white in summer with perhaps a few rills trickling through the gravel bars which shift their position almost every winter.

It is obvious that the geomorphic processes in a Mediterranean climate are far more capable of effective action than those of Central Europe. It is also evident that they are likely to have been affected, especially in the European Mediterranean, by thousands of years of human interference. This is the case with the formation of soil erosion gullies and the various local landslip types (*Frane*), in the accelerated formation of karst by deforestation and also in the increased vigour of torrential streams (with increased surface wash in the mountain areas and increased deposition in the basins and coastal plains, as noted by Philippson, 1947). The very existence of these humanly induced effects appear to add weight to the idea that in Mediterranean areas (like the steppe) there is a much greater *natural* tendency towards greater geomorphic activity, for they are known to be present in the absence of such secondary influences, as in northern Chile (Mortensen, 1927). Finally, many of the especially characteristic features of the European Mediterranean area, such as the larger karst landforms, are obviously much older than all human influences, and must not be overlooked.

All these processes and features, then, are typical of the Mediterranean light forest and shrub areas, for example the sclerophyll areas of the *maquis* and *garigues* in the European Mediterranean. On the one hand, analogies with the temperate mature soil zone and even with the frost debris and desert-like dry debris zone are to be found while, elsewhere, similarities to the typical landforms of tropical savanna and monsoon climates may be seen, that is in the inselberg landscape. In the Mediterranean steppe the periods of highest rainfall occur in the hot season as they do in the savanna and monsoon lands. There are two contrasting examples of the Mediterranean version of the inselberg landscape which might be cited here. First, in the coastal Mediterranean steppe (from about the southern foot of the Atlas of Morocco to Libya) there is the expectable winter rainfall maximum associated with fairly high temperatures. In the continental high-steppe regions, by way of contrast (Central Anatolia, Armenia and north-west Iran) it is sudden storms in *summer* which give rise to sheet-flood activity (see Jaranoff, (1942) on south Morocco, Rathjens senior (1928) on Libya, Louis (1949) on Inner Anatolia, and Bobek (1950) on north-west Iran). Bobek, in fact, stressed that such sheet flooding occurred in a zone from the steppe to the edge of the desert.

To summarise then, the Mediterranean zone not only exhibits similarities to neighbouring zones, but also features approaching those of the more distant climatic-morphological zones of the tropics and extratropics. It seems certain that more intensive work will establish the Mediterranean forest as a distinct climatic–morphological zone. On the other hand, whether one should view the various Mediterranean steppe types as Mediterranean subtypes or as belonging to the dry debris sheetwash zone transition, remains for the time being undecided. For this reason, we have written on the map (Fig. 7.1) 'isolated inselbergs' in the steppe border zones of the Mediterranean mountain areas.

Dry debris zone

All deserts exhibit one very marked climatic–morphological characteristic, namely the predominance of mechanical weathering over chemical. This is the direct consequence of a lack of rainfall and plant cover. The 'soil' of all desert areas, therefore, consists mainly of dry debris of all sizes and assortments, from blocks and coarse fanglomerates to dune sand and the finest dust. From the uniformity of their weathering processes arises the similarity between their erosion processes. For the purposes of this initial account, all the world's deserts have been grouped into one large climatic–morphological zone—the dry debris zone (*Trockenschuttzone*). It is similar in total extent to the Penckian 'arid zone', although it does not fit so clearly as the glacier zone fits Penck's 'nival' category. However there are important divergences from Penck's system, especially in the border areas, for Penck's division between full-arid and semi-arid on the basis of rainfall alone is inadequate for the purposes of climatic–morphological classification. As a relative decrease in linear erosion is experienced here, any classification is bound to be heavily dependent on the variable balance which exists between the main denudation processes—insolation weathering, salt shattering, sheetwash, rill-wash, the movement of saturated debris on slopes and deflation.

Apart from Penck's hydrological classification, there has been a long series of desert classifications based partly on climate, partly on tectogenetics and partly on rock type. Most researchers, however, have found it impossible to separate the purely climatic factors from the others and therefore have not succeeded in establishing a generally acceptable classification of deserts. The problem lies in the fact that, with a complete lack of plant cover on almost all the steeper slopes, the parent rock appears at the surface, thus influencing the formation of small features more than in any other zone. Even the debris layer which covers the parent rock on the gentler slopes, however, is composed almost entirely of this same rock so that its distribution varies accordingly. Only after sizeable movements of the debris cover have taken place or after it has broken down into sand and dust is there any appreciable modification of the structural influence on the landforms. The problem of deserts has been overcome most successfully in the tropics. Here Mortensen (1950) has taken the north-Chilean desert as his example and has arranged a purely climatic classification of desert types consisting of concentric circles from the core area outwards. The differences in the efficacy of the predominant formative processes are traceable to the difference in the importance of salt shattering (hydration). In the central desert zone this process is unimportant, so that in a given period of time little loose material is made available for transport. This disintegrates *in situ* into a very fine dust, which is protected from the wind by the presence of that still unresolved phenomenon, a thin crust or 'dust skin' (*Staubhaut*), which has also been observed by Passarge and Meinardus (1933) in the Egyptian desert. In the absence of wind action, landform modification in the heart of the desert is left mainly to the various 'wash' effects of flowing water. This clearly distinguishes the central desert from the other wind-affected deserts, at least qualitatively. The great rarity of morphologically effective precipitation also implies generally weaker geomorphic processes here. In spite of this, it seems indisputable that such areas are still subject to greater erosion intensities than the temperate zone.

On the other hand, in the second major climatic desert type (the peripheral desert (*Randwüste*), in which Mortensen rightly has to incorporate all the remaining desert types), the effects of erosion are altogether greater. The initial attack by salt shattering and insolation is very great here. Even if the fine material is not protected from the wind by a 'dust skin', there can be no doubt whatever that the greatest erosional effects are due to running water. On the steeper slopes it is slope-wash and gully-wash, whilst on the gentle slopes it is sheet-wash and rill-wash which are predominant, leading to the development of gently sloping debris-veneered rock surfaces and inselbergs. The gently sloping surfaces of the deserts and dry savanna are characterised by low-gradient water courses with flat, wide debris-covered floors, quite unlike those in the wet savanna which are rendered morphologically distinct from the surrounding countryside by a sharp break of slope (Obst and Kayser, 1949). It is in this 'peripheral desert' that individual geomorphic processes vary considerably under the influence of local geological (petrographic and structural) conditions so that it is not yet possible to infer a more refined climatic–morphological subdivision. Such a subdivision would at least separate the 'tropical' from the 'extra-tropical', in which relationships between erosional processes are obviously slightly altered due to decrease of surface wash and increase in frost action. This becomes increasingly important in the tropical and extratropical high-altitude deserts, so that still further types must be distinguished (Troll, 1944, 1948(*a*), 1948(*b*)). The vast extratropical high-altitude deserts, in for example Tibet, represent the only case in which the dry debris and frost debris zones border directly on to a forest soil area or temperate mature soil zone.

The desert steppes, the flatter areas of which are surfaced by *orts* soils containing a considerable volume of chemical decomposition products (sierozems, grey soils and tropical earths), can no longer be classed with the dry debris zone, in our opinion. As mentioned above, the position of the extratropical and, in part, the subtropical desert steppes in this scheme must remain undecided. The tropical desert steppes and dry savannas are, on the other hand, clearly a part of the sheetwash zone which is discussed below.

Even without the desert steppes and the exceptions of the central desert areas, there remains in the peripheral deserts a large area which is still to be classified satisfactorily. It is probable that there are important qualitative and quantitative differences to be discovered here. The dry debris zone (as well as the Mediterranean zone) lacks the well-marked geomorphic style to be found in the other areas discussed above.

Sheetwash zone

Apart from the temperate zone of mature soils and the dry debris zone, this is the most extensive of the climatic–morphological zones and incorporates the whole area of the seasonally moist tropics. This vast region lies between the dry debris zone and the inner tropical forest zone and is made up of the tropical desert steppe, and the tropical steppe and savanna lands on the western side of the continents where regular rainfall is experienced. However, geomorphic processes, very similar in their effects, predominate in the monsoon lands on the eastern sides of continents. In fact, they extend right into the subtropical monsoon areas directly bordering the temperate mature soil zone on the eastern side of the inland deserts. In spite of its great extent, the sheetwash zone forms a very distinctive, scarcely subdivided unit which was recognised as such a long time ago (Waibel, 1925, 1928 and Jessen, 1936). Krebs (1942) examined its extent on the basis of the distribution of inselbergs and, using certain climatic statistics, found a spatial, if not a completely unambiguous, functional relationship.

The main characteristic of this zone is the predominance of two unique and efficient denudation processes namely sheetwash and rill-wash. Both processes are closely associated with great sheet-floods arising from regular thunderstorms which occur at the peak of the hot season. The surface erosion thus caused is so powerful that it obliterates the morphological effects of most of the smaller erosion channels, leaving only linear erosion by the larger rivers as in any way comparable. If there is a fair balance between surface and linear erosion in the soil flow zone, and a slight predominance of linear erosion effects in the temperate mature soil zone and Mediterranean zone, then in the sheetwash zone we find a definite predominance of surface denudation processes over linear erosion processes. The predominance of one single denudation process type is unmistakably expressed in the range of landforms—almost as clearly, in fact, as in the dry debris zone. The present-day sheetwash zone thus probably represents an extreme case. How closely it compares with the frost debris zone in erosional capacity is difficult to say, but in the individuality of its characteristic landforms (the extensive rock pediments and stepped erosional plains or *Rumpftreppen*, with the sharply defined inselbergs), it is perhaps the outstanding subaerial zone containing, next to the glacier zone, the clearest morphological features on the earth's surface.

The combination of very powerful chemical weathering with especially effective sheetwash and rill-wash sharply emphasises bedrock variations in contrast to most of the dry debris zone. Even erosional scarp development is to some extent retarded as I have suggested earlier (Büdel, 1938), a point proved by Obst and Kayser (1949) in their work on the plateau scarp in eastern South Africa. That other major example of geologically determined landforms, the karst, here exhibits 'dome karst' characteristics, as H. Lehmann (1936) has described it in Java. As these approximate to inselberg form, such areas may be included in the term 'inselberg landscape'. Outside the large, old continental blocks which exhibit the inselberg landscape most vividly, this dome karst landscape extends often in somewhat altered form, even into young-fold mountains which, in this zone in particular, tend to develop unduly broad valley bottoms and very flat intramontane plains (Credner, 1935).

The mature soil zone of the inner tropics

This last zone possesses, in somewhat exaggerated form, many of the climatic–morphological characteristics of the continually wet temperate mature soil zone. The continually high temperature and humidity, and the luxuriance of the primeval tropical forest, have led to the formation of a very deeply leached soil cover due almost entirely to chemical weathering in an area where the climate has remained constant through long periods of earth history. Because of the thick soil and the forest cover, the underlying rock is completely protected from direct climatic influence.

On all the more gentle slopes, the effects of denudation processes are normally restricted to the thin, uppermost layer of the soil cover—a situation similar to the corresponding temperate zone. With increasing slope, there occurs a transition to more powerful erosion processes in the form of subsurface earth flows and slips (Behrmann, 1915, 1927; Sapper, 1935) which continually sharpen the ridge forms, especially in the higher mountain areas. However, these processes do not prevent the underlying rocks on even the steepest slopes (up to 55°) from being covered by a thick chemically weathered soil and a forest cover. For this reason, most of the rivers are free of coarse debris and therefore tend to have steep-sided valleys, caused by powerful vertical erosion. Accordingly, as Behrmann pointed out, there is a relative predominance of linear erosion processes over surface denudation processes: a similar balance to that found in the temperate mature soil zone and the Mediterranean zone. It seems certain that further research will lead to further subdivisions, although the internal differences may prove to be not very marked. In quantitative terms, the erosive capacity of this zone cannot be very great, to judge by the small amount of debris produced, and the ability to produce distinctive landforms is considerably less than is found in the sheet-wash zone.

CLIMATIC–MORPHOLOGICAL ANALYSIS OF MAJOR LAND-FORMS

Now we come to our second major task—the climatic–morphological analysis of the major landforms themselves. This is undoubtedly the main aim of climatic geomorphology. The difficulty, as has already been stressed many times, lies in the fact that in many parts of the world such rapid climatic change has occurred in the recent past that, next to the landforms that have been formed by presently active processes, there are still numerous 'paleo-landforms' reflecting former climatic–morphological relationships. The question is: where are the former predominant and where the latter?

Paleo-landforms

Naturally we least expect to find paleo-landforms in areas where current processes are extremely powerful, and vice versa. Perhaps the simplest situation in which to test this relationship is one where paleo-landforms have developed in a large region consisting of several zones and where the change to the present climatic conditions took place simultaneously throughout. Only in such a situation will the more powerful processes reflect the overriding influence of the major landforms; and only then can the degree of this influence on the efficacy of current processes be assessed. Clearly, even weak processes can, given enough time, impress their characteristics on the crust, just as more powerful processes do in a shorter time.

In fact, such requirements have been fulfilled to a varying degree over a large part of the earth. The reason for this is the Ice Age. To put it more precisely, periods of glaciation caused long-lasting climatic changes throughout many parts of the world, such as they had not experienced during the whole of the Tertiary and the Mesozoic periods, that is over 200 million years. Moreover the Ice Age lasted so long (at least several hundred thousands of years, the final glaciation alone lasting 50–100 000 years) that the face of the earth was extensively altered by the geomorphic processes. This new type of landscape was fundamentally different from the landscape which preceded and succeeded it. This landscape system came to an end simultaneously all over the world about 12 000 years ago at the end of the Ice Age. This point of time is geologically so recent that only in areas where very active processes are operative has there been a marked alteration of the glacial and associated landforms. In zones where the processes are less active the glacial features are preserved almost unaltered.

Thus we have, in fact, a temporally and qualitatively uniform basis for assessing the impact of current processes on the landscape. This enables us to do three things: first, to establish with fair certainty the proportion of currently active features in the landscape in several zones; second, to gain exact quantitative measurements of erosion capacity per unit of time in individual zones; and, finally, it allows us to recognise qualitatively a zone's landform character, at least in areas which have developed their own landform 'style' since the Ice Age.

While it is true that the end of the last glacial period and the beginning of the present climatic régime occurred simultaneously throughout the world, the climate of the glaciations did not produce the same landforms everywhere. Thus the second condition of our observations is not fulfilled so well over all parts of the earth's surface. In order to be on safer ground a new reconstruction of the Pleistocene climatic belts has been undertaken (Büdel, 1949). This shows that at that time all the climatic belts were in a position different from those of today. One striking difference is in the location and extent of the extratropical and tropical zones. The latitudinal shift of the climatic belts was at its greatest near the poles, so that the southern limits of the glacier, frost debris and tundra zones were at that time 22°C to 24°C more southerly than they are today. Towards the equator the shift in climates was less; the southern limit of the temperate zone of mature soils was about 9°C, that of the Mediterranean zone about 5°C more to the south than today. In the tropics the shift was even less, in fact a slight poleward movement is detectable in the case of the inner tropical leached soil and the sheetwash zones, so that the dry debris zone (O. Selling, 1948) was constricted (the core areas of the tropical deserts being appreciably wetter than they are today). One result of all this is that the present-day tropical climatic belts, as opposed to those of the Pleistocene, have experienced only a small peripheral shift, the core areas remaining fixed. This is why the core areas of the inner tropical leached soil zone and the sheetwash zone displayed almost the same morphological character then as now. On the other hand, the subaerial belts outside the tropics (from the edge of the glacier zone to the central part of the Mediterranean zone, inclusive) have moved a long way polewards since the end of the Ice Age. They now occupy zones which were formerly occupied by completely different climatic zones. In these cases, the development of the present landforms began simultaneously everywhere, on the variety of 'morphological datum planes' presented by the different zones. For three of these zones (the present frost debris zone, the tundra and the greater part of the temperate mature soil zone), this Pleistocene picture was exactly the same: all three zones now lie in the area which was formerly covered by vast ice sheets. It is here, then, that the best conditions exist for the recognition of differences in morphological character from one climatic zone to the next. Accordingly, the key to the main problem of climatic geomorphology lies outside the tropics.

Pleistocene features

For these three climatic–morphological zones which coincide with the Pleistocene glacier zone, Büdel's analysis (1948) shows that in the frost debris zone almost all the landscape features of the former glacier zone, except for a few remnants preserved under postglacial ice or sea or near especially hard rocks, have been obliterated and that characteristic present-day features have been imposed. In the short time since the Ice Age, the frost debris zone has demonstrated its considerable erosive capacity. On the other hand the efficacy of erosion in the tundra zone since the last glaciation has been much less. True, the unconsolidated landforms of glacial deposition have been remodelled to a great extent and the finer features of glacial erosion have been incorporated into some postglacial features, but underneath this superficial superimposition of forms the Pleistocene legacy to the landscape is still clearly recognisable. In the temperate zone of mature soils, on the other hand, alteration of glacial features, outside the major river courses, is largely absent. The effect of post-glacial denudation has been so small that the greater part of the land surface and even the old glacial soil layers are still preserved. In contrast to the frost debris zone, then, glacial features have been preserved in all their morphological detail.

This is true, also, of those features left behind outside the limits of the former ice sheet. These have great morphological individuality, consisting of frost debris and glacial tundra zones which formerly extended to the northern edge of the present-day Mediterranean area, that is as far as the Pyrénées, the southern edge of the Massif Central, the Alps and the Dinaric Alps. Between this line and the edge of the northern ice sheet this zone was far more extensive than the arctic of the present day. Climatic–morphological analysis of this well-preserved tundra zone has shown that, despite many similarities with its present-day equivalent, it possessed a somewhat different character and also a richer diversity. Its more southerly latitude gave it a tundra climate that was warmer in summer and more continental in character, than is true of any tundra areas of today from Iceland to eastern Siberia. For this reason the effect of wind in the Pleistocene tundra zone played a far more important role than it does today. The south-eastern part of the Pleistocene tundra zone looked like the so-called loess steppe. This was not forested but, because of the higher summer temperatures it did not altogether lack trees, and in places where there was greater moisture, forest tundra or forest-steppe existed. In spite of some differences, however, the remains of these features show that the geomorphic processes in the Pleistocene frost debris and tundra zones were so similar to those of the present that we can use the same terminology for both. Considering the major landforms produced, the main difference between the two seems to be that processes were operative for a much longer period (several hundreds of thousands of years) in the Pleistocene frost debris and tundra zones so that the suite of landforms produced was more complete.

Thus, in the present-day temperate mature soil zone, all Pleistocene features of the glacier, frost debris and tundra zones are equally well preserved. As this Pleistocene tundra zone extended southward into this zone only periodically, it is today full of fossil periglacial features. However, there is a close correlation between the extent of the existing frost debris and tundra zones and the distribution of currently active typical landforms. In the temperate zone, on the other hand, classification of the existing landscape on the basis of current processes rather than on those of the glacial period, is impossible. We can still distinguish in this region glaciated landscapes, young and old moraine areas, periglacial tundra zones, loess steppes, etc. Accordingly, we can assess the efficacy of current processes in this zone only by process study and certainly not by interpretation of the existing landforms.

In contrast to the climatic–morphological zones with such slight erosive capacity, the more active zones serve as a geomorphic yardstick (*Prägstöcke*). Their influence, as we have seen in Pleistocene Europe, extends far beyond their present area of operation. Once they have put their stamp clearly upon an area, their traces cannot be quickly obliterated by weaker processes resulting from climatic change. Given relatively rapid climatic change, such as has occurred since the end of the Tertiary, these controlling characteristics remain dominant during periods of weak geomorphic activity until they are eventually overrun by a new set of controlling characteristics. Such ancient landform remnants are naturally preserved longest well away from the main lines of erosion, that is in the watershed areas of tectonically quiet regions. On the high plateau areas of the Mittelgebirge such erosion surfaces and inselbergs are widely preserved. These are relics of a Tertiary sheetwash zone which persisted despite the prevalence of tundra zone conditions for a period of several hundreds of thousands of years during the Pleistocene. Certain individual characteristics have been radically altered, but as a whole the features remain vary well preserved. Even under an ice cover they remain preserved in the Kjölen area, and they are also to be found in Spitzbergen, which now lies in the frost debris zone. From this it is obvious that, next to the glacier and frost debris zones, the sheetwash zone represents an especially important climatic–geomorphic indicator. Whether it exceeds the glacier and frost debris zones in absolute erosive potential remains doubtful, however; first, because the operative climate has prevailed far longer than have the glacial climatic zones, and, second, because in creating extensive rock-cut erosion surfaces it has etched into the earth's surface a landform element which is extremely difficult to eradicate.

Current processes in tropical and subtropical landscapes

Finally, let us take a brief look at the effect of present-day processes in tropical and subtropical landscapes where, as has already been said, we unfortunately lack a frame of reference to help us gauge the extent of present-day influences. Even in the Mediterranean transition zone, the existence of a complex of Pleistocene remnants cannot be demonstrated. One reason for this is the vigour of current processes. But, more important, it is probable that geomorphic processes were far weaker during the Pleistocene cool phases than, for example, in Central Europe for, according to paleo-botanical research, the Mediterranean transition zone appears to have been similar to the temperate mixed forest belt during the Pleistocene glacials. We have already noted the lack of broad floodplains of *Würm* age in the Mediterranean valleys which may be taken as some corroboration of this view. The characteristic features of the zone were not altered very much during the Ice Age. On the contrary, the processes were interrupted by periods of calm which tended to preserve them. The contrast in features, then, between this zone and the temperate mature soil zone rests not only on the completely different activity evident under the present climate but also on the contrast in processes during the Pleistocene. The climatic–morphological individuality of the Mediterranean transition zone is thus further re-emphasised.

On the northern edge of the dry debris zone, on the other hand, we again find paleo-features, particularly those due to valley formation in the Pleistocene pluvial periods. In the interior of the dry debris zone the climate was markedly wetter than it is today. Its climatic–morphological subzones were not completely replaced in the Pleistocene glacials; rather they possessed traits similar to those of today, and only suffered slight peripheral shifts.

The inner tropical zone of mature soils has been more stable than the other zones for a long time. As in the world of flora and fauna, so in that of climatically determined morphogenesis it remains, despite the Ice Ages, a vast, unchanged reminder of Tertiary conditions. A satisfactory geomorphic analysis of the zone is still awaited. It should prove especially important because it may constitute a broad analogue of the landscape which our temperate mature soil zone is striving to achieve, were it to remain in these latitudes and under the same climate for a lengthy period of time.

REFERENCES

BEHRMANN, W. (1915). Der Sepik und sein Stromgebiet. *Erganzungsnr*. **12**, Mitt. Dt. Schutzgeb

—— (1927). Die Oberflachenformen im feuchtheissen Kalmenklima. *Dusseldf. Geogr. Vortrage u. Erortg.*

BOBEK, H. (1950). Vorgange und Formen der Abtragung in Nordwestiran.

BÜDEL, J. (1938). Das Verhaltnis von Rumpftreppen zu Schichtstufen in ihrer Entwicklung seit dem Alttertiar. *Pet. Mitt.*

—— (1944). Die morphologischen Wirkungen des Eiszeitklimas im gletscherfreien Gebiet. *Geol. Rdsch.* **34**, Beitrage 1

—— (1948). Die klimamorphologischen Zonen der Polarländer. *Erdk*. **2**, 1–3, Beitrage 2

—— (1949). Die raumliche und zeitliche Gliederung des Eiszeitklimas. *Die Naturwiss.* 36 Beitrage 3

—— (1950). Die Karstgenerationen der Schwabischen Alb und der Nordostalpen.

CREDNER, W. (1935). *Siam, das Land der Thai.* Stuttgart

JARANOFF, D. (1942). Ergebnisse einer Forschungsreise nach Nord- und Westafrica. *Forschg. u. Fortschr.*

JESSEN, O. (1936). *Reisen und Forschungen in Angola.* Berlin

KÖPPEN, W. (1936). *Das geographische System der Klimate.* Handb. d. Klimatologie, hrsg. v. W. Koppen u. R. Geiger, Bd. I Teil C. Berlin

KREBS, N. (1942). Uber Wesen und Gestaltung der tropischen Inselberge. Abh. Preuss. *Ak. d. Wiss., Math.-Nat.* Kl. 6

LEHMANN, H. (1936). Morphologische Studien auf Java. *Geogr. Abhdl.* 3 H.9 Stuttgart

LOUIS, H. (1938). Bemerkungen über den Bewegungsmechanismus der Gletscher und Folgerungen daraus fur die Theorie der Glazialerosion. *Comptes rendus du Congrès Int. d. Geogr. Amsterdam.* Leiden

—— (1949). Anatolien. Unveroffentl. *Vortrag Geogr. Kolloqu.* Gottingen

MENSCHING, H. (1950). Schotterfluren und Talauen im Niedersachsischen Bergland. *Gott, Geogr. Abhandl.*, H.4

MORTENSEN, H. (1927). Die Oberflachenformen der Winterregengebiete. *Dusseldf. Geogr. Vortrage u. Erortg.* 3

—— (1930). Einige Oberflachenformen in Chile und auf Spitzbergen im Rahmen einer vergleichenden Morphologie der Klimazonen. *Pet. Mitt.* Erg. 209

—— (1949). Rumpfflache-Stufenlandschaft-Alternierende Abtragung. *Pet. Mitt.*

—— (1950). Das Gesetz der Wustenbildung. Universitas, 5.

OBST, E. and KAYSER, K. (1949). Die grosse Randstufe auf der Ostseite Sudafrikas und ihr Vorland. *Geogr. Ges.* Hannover

PASSARGE, S. and MEINHARDUS, W. (1933). *Studien in der ägyptischen Wuste.* Berlin

PENCK, A. (1910). Versuch einer Klimaklassifikation auf physiographischer Grundlage. *Sitz-Ber. Preuss. Ak. d. Wiss. Phys.-Math.* Kl, **12**

PHILIPPSON, A. (1947). *Land und See der Griechen.* Bonn

RATHJENS, C. sen. (1928). Löss in Tripolitanien. *Zt. Ges. f. Erdk.* Berlin
SAPPER, K. (1935). Geomorphologie der feuchten Tropen. *Geogr. Schr.*, hrsg. v. A. Hettner, H.7
SCHMIDT, W. F. (1948). Die Steppenschluchten Sudrusslands. *Erdk.* 4–6
SELLING, O. H. (1948). On the late Quaternary history of the Hawaiian vegetation. Honolulu, Hawaii
TROLL, C. (1944). Strukturboden, Solifluktion und Frostklimate der Erde. *Geol. Rdsch.* 34/8/8
—— (1948a). Die Formen der Solifluktion und die periglaziale Bodenabtragung. *Erdk.*, Bd. I
—— (1948b). Der asymmetrische Aufbau der Vegetationszonen und Vegetationsstufen auf der Nord- und Sudhalbkugel. *Ber. geobot. Forsch. Inst. Rubel in Zurich fur* 1947
WAIBEL, L. (1925). Gebirgsbau und Oberflachengestalt der Karrasberge in Sudwestafrika. *Mitt. d. Dt. Schutzgeb.* 33 Berlin
—— (1928). Die Inselberglandschaft von Arizona und Sonora. *Ztschr. d. Ges. f. Erdk.* Sonderband

8 The Geographic Cycle in Periglacial Regions as it is Related to Climatic Geomorphology

LOUIS C. PELTIER

DAVIS, in his early writings (1899–1909) recognised, by implication, that some aspects of the geographic cycle were related to the climatic régime under which they were formed. This was a corollary to his classic formula: form is the result of the influence of structure plus process plus stage. Davis, in recognising the existence of the arid and glacial cycles, which he called climatic accidents, also recognised that process was dependent on climate.

This implication was examined in greater detail by various European writers, notably Albrecht Penck (1905, 1910), Passarge (1926) and Thorbecke (1927), and in America by Bryan (1940). Recently Büdel (1944, 1948) has suggested the recognition of *formkreisen* or *morphogenetic regions*. If the geographic cycle were defined in terms of the climatic régime under which it develops, the minor forms of the landscape would become emphasised. Regional distinctions in topography which transcend structural regions might then become defined in our consciousness. Even changes in climate would become recognisable in the landscape.

It is not possible, with our present knowledge, to formulate precise definitions of the morphogenetic regions in terms of climate or the peculiar erosion cycle with which they are to be associated. One can only speak qualitatively and on the basis of personal impression. Nine different climatic régimes of geomorphic significance, and therefore nine different morphogenetic regions may, however, be tentatively postulated. Their recognition is based upon the supposed relative significance of the different geomorphic agents in a given climatic region. The climatic distinctions used here are crude and might be considerably refined. As, however, our knowledge of the relative importance of the geomorphic agents is even less precise, the climatic régimes are presented in terms of mean annual temperature and rainfall only. See Table 8.1: this form of climatic diagram resembles those used by Blumenstock and Thornthwaite (1941) and also many of their predecessors.

Table 8.1 Morphogenetic regions

Morpho-genetic region	Estimated range of average annual temperature	Estimated range of average annual rainfall	Morphologic characteristics
	°C	cm	
Glacial	−18/−7	0/114	glacial erosion nivation wind action
Periglacial	−15/−1	13/140	strong mass movement moderate-to-strong wind action weak effect of running water
Boreal	−9/3	25/152	moderate frost action moderate-to-slight wind action moderate effect of running water
Maritime	2/21	127/190	strong mass movement moderate-to-strong action of running water
Selva	16/29	140/229	strong mass movement slight effect of slope wash no wind action
Moderate	3/29	89/152	maximum effect of running water moderate mass movement frost action slight in colder part of the region no significant wind action except on coasts
Savanna	−12/29	64/127	strong-to-weak action of running water moderate wind action
Semi-arid	2/29	25/64	strong wind action moderate-to-strong action of running water
Arid	13/29	0/38	strong wind action slight action of running water and mass movement

Seven of the morphogenetic regions roughly coincide with 'geographic cycles' or 'climatic accidents' previously described. Two other climatic régimes should also have geomorphic processes of differing character or intensity and should therefore produce characteristic geomorphic features. They are the *boreal* and *marine west coast* climates described by Köppen (1923) as the Dfc and Cfb climates respectively.

That the processes of intensive frost action in the periglacial areas constitute a cycle has only recently been recognised by Bryan (1946) and Troll (1948). As this cycle is current only in remote portions of the arctic and in high mountains, but was of great importance in the periglacial areas of temperate regions during the Pleistocene, it is here called the periglacial cycle following Troll (1948). In Davis's terminology the periglacial cycle is a 'climatic accident'. If, however, the régime of the periglacial cycle be considered in comparison with that of the other cycles, including the moderate or 'normal' cycle (Davis's) it may be found to have equally persistent characteristics. The cycle must then be supposed to have equal validity with other cycles.

Each cycle is here considered to be 'normal' within its own régime and a 'climatic accident' only when it temporarily encroaches upon the area of another régime. Any climatic régime may produce both a normal cycle and a climatic accident, depending upon the fluctuations of climate that may have occurred in the past. For these reasons the name 'moderate cycle', which implies origin under a climate of moderate temperatures and rainfall, is used here for the cycle of moderately humid, temperate regions, in place of 'normal cycle'. Lands which fall within the zone of climatic fluctuations may have the peculiar characteristics of one cycle superimposed upon those of another. A composite product of the two régimes, here called polygenetic topography, would then result. In a consideration of the effect of human activity on erosive process and in the analysis and description of the landscape, the concepts of morphogenetic regions and polygenetic topographies may be useful.

THE GEOGRAPHIC CYCLE AND MORPHOGENETIC REGIONS

If Davis's line of investigation be followed in the direction indicated by Albrecht Penck (1905, 1910), Hettner (1921), Passarge (1926), Thorbecke and others (1927), Bryan (1940), Blumenstock and Thornthwaite (1941), Büdel (1944, 1948), Troll (1948) and others, a series of climatic régimes may be established within which the intensity and relative significance of the various geomorphic processes are, according to our present information, essentially uniform.

These regions are morphogenetic regions. They serve to further define and describe the geographic cycle.

A distribution of morphogenetic regions, with respect to either climate or space, may be postulated from a consideration of the realms of activity of the various geomorphic processes. These realms may be further subdivided by considering whether the process is merely present, or is dominant or intermediate. These hypothetical realms and the morphogenetic regions postulated therefrom are presented below on graphs in Fig. 8.1. Because, as has already been explained, accurate means of measuring and comparing these processes are not available, these graphs represent merely a diagrammatic exposition of a concept and are necessarily a reflection of the author's reading and thought. Inevitably the ideas of others have been used and the author here acknowledges his debt to his predecessors whose writings have inspired him.

The different morphogenetic elements consist of the processes of rock weathering and the transportation of these products. The processes of rock weathering consist of chemical decomposition and mechanical disintegration. Together they account for the major part of the material made available for transportation. The ratio of their effectiveness to the effectiveness of transportation determines the presence or absence of exposed rock and the thickness of residual soils.

Chemical decomposition, which consists primarily of the oxidation, hydration and solution of the various mineral constituents of the rocks, may be theoretically related to the climate. Rainfall, in so far as it determines the availability of water for chemical reactions, and temperature, as it determines the speed of chemical reactions—these are essential elements. All the processes of chemical decomposition require water either as a constituent in the reaction or as a solvent and transporting agent for the products of the reaction. The presence or absence of water is broadly synonymous with the presence or absence of chemical decomposition. If it is assumed that the availability of water varies directly with the rainfall, then, other conditions being assumed equal, the probable rate of chemical decomposition will vary directly with rainfall. Furthermore the rate of chemical reactions increases as temperature increases (Arrhenius, 1889) which may, however, be offset by a distinct decrease of solubility of oxygen and carbon dioxide in water with increased temperature (Glasstone, 1940). This direct variation of chemical weathering with temperature is augmented in humid lands by the increased density of vegetation, with increase in both rainfall and temperature, leading thereby to an increased production

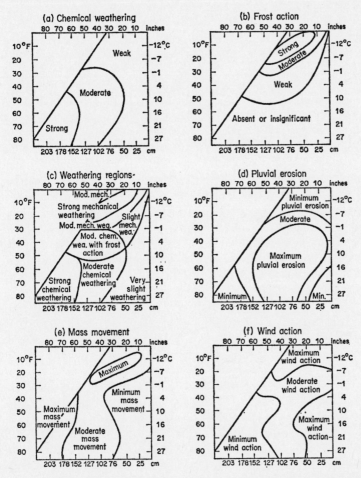

Fig. 8.1 Hypothetical realms and morphogenetic regions: mean annual rainfall and temperature

of organic acids in the soil. Chemical reactions which are related to organic wastes are, therefore, most rapid in the warm, humid regions.

The relations set forth above form a crude but effective measure of the relation of chemical decomposition of rocks to climate. In

Fig. 8.1(a), the maximum rate of chemical decomposition is placed in the humid tropics and there are shown two zones of minimum rate of decomposition, one in the low-latitude deserts and the other in the cold, high-latitude deserts.

Mechanical disintegration, where it is divorced from agents of transportation so as to exclude abrasion and corrasion, probably is largely frost produced. The idea that mechanical weathering or the break-up of rocks by temperature changes is characteristic of desert areas has largely been abandoned under the attack by Black-welder (1925). For this reason frost action is taken, in the following discussion, to represent the overwhelmingly greater part of mechanical distintegration. As has already been amply shown by Steche (1933) and others, the intensity of frost action is dependent upon the frequency of temperature fluctuations about the freezing-point in the presence of water. There are, thus, two thermally-controlled minima of frost action, where the temperatures are too warm for freezing and too cold for thawing. The final picture of the distribution of frost action is derived by superimposing the rainfall pattern with its subtropical and arctic minima upon the thermally controlled pattern. The result, shown in Fig. 8.1(b), indicates that there is a maximum zone of frost action located in the relatively humid subarctic regions. Frost action decreases in intensity not only in the direction of the warmer and drier regions, but also toward the colder and drier areas of the far north.

The distributions of chemical decomposition and of mechanical disintegration, as represented by frost action, shown in Figs. 8.1(a) and (b), may be combined to produce a generalised representation of the distribution of weathering as shown in Fig. 8.1(c). This generalisation indicates that there are seven major categories or regions in which weathering processes take characteristic forms. As these processes produce soils they may be considered also as pedogenetic regions. Within these pedogenetic regions not one of the weathering processes occurs exclusively. In each region the different processes occur together in unique proportions.

Rapid chemical weathering prevails in the humid tropics, rapid mechanical weathering in the humid subarctic. In the arid regions the rate of weathering of any kind is relatively slight. Moderately rapid chemical decomposition probably prevails in the semi-arid low latitudes, and slow or relatively insignificant frost action may be anticipated in the semi-arid subarctic regions. Both chemical and mechanical weathering are, in this diagram, assumed to be present in the humid, mid-latitude regions. If the rate of erosion and erosional history are assumed to be uniform and constant, the deepest residual soils should occur in the humid tropics and the humid subarctic, and the shallowest residual soils in the mid-latitude and high-latitude deserts.

AGENTS OF TRANSPORTATION

The comminuted material provided by these weathering processes is subject to erosion by one or more of three agents of transportation. These are running water, mass movement and wind. The action of waves of the sea is here excluded from consideration, for its activity, although universal in all climates, is limited in area to the coasts. Waves cannot have the same significance in the development of morphogenetic regions as do the other three agents of transportation. The geographic cycle of wave action or marine cycle (Johnson, 1919) has, however, a certain validity. Its local peculiarities and possible subdivisions related to the climatic characteristics of storminess, wind variability and the biologic elements of the growth of grasses, forests or corals lie beyond the scope of Johnson's marine cycle. When the marine cycle has been considered from a climatic standpoint it may possibly be recognised in relation to polygenetic topographies.

Running water

The importance of running water as a geomorphic agent was emphasised by Davis. Within the regions in which he made his observations it is certainly the most effective of the agents. Running water, as it is considered here, includes slopewash and rill-wash as well as the flow of both clear and turbid streams. This agent cannot be everywhere equally important in every landscape, for its action is dependent upon five main variables: the intensity of the precipitation, the frequency of storms, the permeability of the ground, the rate of evaporation and transpiration, and the nature of the plant cover. Also, in order to permit the maximum effectiveness of fluvial erosion, the rate of weathering must equal or exceed the rate of removal by water. Under these circumstances barren rock would never be exposed and the running water would always be flowing over comminuted and readily transported material. The effectiveness of running water as an erosive agent is further limited by the vegetation. Wherever the plant cover is sparse or only moderately dense the vegetation acts to inhibit gully erosion, but does not prevent slope-wash and rill erosion. On the other hand, wherever the plant cover is dense and the mantle of fallen leaves lies thickly upon the floor of the forest, the leaves act as a roof to prevent the infiltration of rain-water into the ground. The water tends to run off on the surface of the fallen leaves and both slopewash and gullying may nearly cease. This situation is marked in the tropics and was observed by the writer in 1944 and 1945 on the flanks of the northern end of the Cykloop Range near Hollandia and Tanah Merah in New Guinea.

Considering the foregoing all too brief résumé of the principles of erosion, three areas of minimum effectiveness or speed of erosion by running water may be outlined. These, shown in Fig. 8.1(d), are in: the mid-latitude desert regions of low rainfall, the arctic and sub-arctic regions of low rainfall, and the tropical regions of heavy plant cover. The realm of maximum pluvial erosion is similarly shown to be located in a climate intermediate in character between those of the three erosional minima.

Mass movement

Mass movement, as an erosive agent, is here taken to include land-sliding, mud flowage, creep, congeliturbation and solifluction in so far as one differs from the other. All forms of mass movement are produced by the saturation of thick deposits of loose material, sometimes aided by frost action, with the possible exception of dry creep.

The word 'creep' as defined by Sharpe (1938) is so broadly applied as to indicate only the existence of minor or conditional instability and the gradual subsidence of the slope. It cannot be used to relate a process to a climatic régime. Thus creep under frost conditions is congeliturbation; under moist conditions, soil flowage; under dry conditions it constitutes a particular form of displacement brought about by thermal expansion and contraction. No separate term exists for this latter process. It is thus suggested that the word creep be restricted to the process of downslope movement of particles by expansion and contraction through temperature changes in the dry state. Creep may then become a characteristic feature of arid régimes, as contrasted with periglacial and humid régimes.

Two broad categories of mass movement may indeed be recognised, based upon the presence or absence of freezing. The requisite conditions for this process are a rate of weathering which exceeds the rate of erosion (denudation), and an accumulation of soil moisture. The most likely places for the occurrence of strong, or statistically rapid, mass movement are in those parts of the regions of rapid chemical and mechanical weathering, illustrated in Fig. 8.1(c), in which denudation from slopes by sheetwash and gullying is least effective. Two areas of maximum intensity of mass movement, shown in Fig. 8.1(e), may therefore be anticipated, one in the humid tropics (as described by Sapper, 1935), one in the region of marine west coast climate, and another in the periglacial area of intense frost shattering, solifluction and congeliturbation (Kessler, 1925; Steche, 1933; Büdel, 1937; Troll, 1944). Mass movement should be least effective, or statistically slowest, in regions where both the rate of weathering and the available soil moisture are least, in other words the arid and semi-arid regions. Occasional spectacular landslides and mud flows do occur in these regions (as described by Alter, 1926). Many of the landslide scars and deposits of arid regions, however, are ancient and were formed during the moister periods of the Pleistocene (Hoots, 1930; Smith, 1936; Reiche, 1937).

Wind action

Wind action is probably the slowest and least effective of the erosive agents (see Cotton, 1942). This may be true because deflation is limited to grains of the smallest diameter and because deflation does not tend to be as narrowly concentrated at a point or line as are the other, previously considered, agents. Furthermore, minor topographic irregularities and the irregular surfaces of trees and shrubs produce a sufficient roughness to protect the surface of the ground from wind scour. Because of its relatively slight effectiveness, wind action may be apparent only where the effective action of running water and mass movement is slight; because a plant cover reduces surface wind velocities and inhibits deflation, wind action may be apparent only in barren regions; because topographic roughness also inhibits wind velocities, strong wind action should be limited to relatively broad, open country. In a broadly regional sense, excluding topographic influences as well as the influences of coast lines, two areas of maximum effectiveness of wind action may be postulated, one in the warm, dry, mid-latitude deserts and the other in the cold, dry, circumpolar deserts. It is, however, unlikely, except in instances of extreme aridity, that wind action should predominate over the other geomorphic agents, even in these places. The area of least effective wind action is probably that of the humid tropics where the encroachment of swamp vegetation onto the shoreline prevents even the development of coastal dunes. This distribution of wind action is illustrated in Fig. 8.1(f).

Nine different climatic and possible morphogenetic régimes, illustrated graphically in Fig. 8.2, may be postulated from the foregoing analysis. Each should be distinguished by a characteristic

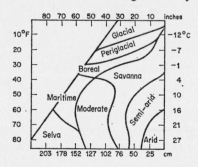

Fig. 8.2   Morphogenetic regions: mean annual rainfall and temperature

assemblage of geomorphic processes. Thus, if cycles are to be defined in terms of the agents and régimes which produce them, they should correspond to nine different geographic cycles. Davis has admitted the existence of the moderate or 'normal' cycle (1899), the arid cycle (1905), and the glacial cycle (1909). Each of these cycles can easily be identified with one of these climatic, morphogenetic régimes. Cotton has recognised a savanna cycle and a semi-arid cycle (1942) which, in a climatic sense, are intermediate between the arid and moderate cycles. He also mentions briefly a 'hot-humid' cycle based on observations by Sapper (1935) and Freise (1938). In order to maintain brevity of name, and in keeping with Cotton's use of the ecological term 'savanna' to describe characteristic processes of the 'inselberg landscape' of Bornhardt (1900), this hot-humid cycle is referred to as the 'selva cycle'. This term is appropriate because the protective effects of the high tropical forest, or selva, prevent or inhibit erosion on hill slopes by running water either as slopewash or in gullies. A thick soil, vulnerable to mass movement, is therefore able to develop (Wentworth, 1943) and the characteristic geomorphic features of this cycle may thereby be formed. The most recently described separate régime is called the periglacial cycle by Troll (1948) and includes the cycle of cryoplanation of Bryan (1946). It will be discussed in more detail below. Thus, of the nine climatic régimes, and corresponding cycles, here postulated, seven have already been described as producing unique geomorphic results. For the remaining two régimes, here called the maritime and boreal régimes, no distinct geomorphic characteristics have so far been reported. The probable erosional features of these as well as the other régimes are presented in Table 8.1. Possibly the unique properties of the boreal régime have been obscured by the relatively recent prevalence of the periglacial cycle. Its morphogenetic characteristics may not be independently discernible. The maritime régime, though in many places recently subject to the effects of the glacial and periglacial cycles, may be recognisable by the severity of mass movement. The limiting climatic boundaries of these régimes are shown graphically in Fig. 8.2. The climatic boundaries of this graph are in part the same as those given by Penck (1910), Davis (1912) and Troll (1947) whose influence is acknowledged by the writer. Particularly parallel to the glacial, humid and arid climates of both Davis and Troll are the glacial, selva and arid régimes of Fig. 8.2. I should, however, say that many of the ideas here expressed are explicitly stated in the lectures of Professor Kirk Bryan, whose emphasis on climatic morphogeny has led to the formulation set forth here.

If the geographic cycle is considered to be the sequence of events leading to the complete destruction of the geomorphic landscape, irrespective of the details of topographic form which may be produced or the details of the geomorphic processes which are effective, there can be but one geographic cycle. From such a broad viewpoint the other cycles here defined are unnecessary. Because the runoff of water from the land surface is, over a long period of time, likely to be the dominant agent in moulding the surface, this single cycle is best described as the *pluviofluvial cycle*. However, if the development in youth and maturity of minor topographic forms, particularly the development of slopes, is to be stressed, then the pluviofluvial cycle is an inadequate framework for the description of surface features peculiar to the several morphogenetic regions. Each geographic cycle is here considered as the unique expression of a climatic régime. Each acts through successive stages upon rocks of differing lithologies which are themselves arranged in various structures.

Landforms are therefore to be described by an expanded Davisian system as structure including lithology, process as modified in nine morphogenetic, climatic régimes, and also stage.

THE PERIGLACIAL CYCLE

One of these cycles, the periglacial cycle, has been described and its properties established by many workers whose reports have appeared throughout the past fifty years. Notable among these are Matthes, Anderson, von Lozinski, Högbom, Cairnes and Bryan. Matthes (1900), while studying the geomorphology of the Bighorn Mountains, recognised *nivation* as a distinct geomorphic process. In 1907, Anderson described the movement of rubble on the slopes of the Falkland Islands. He described the process of movement in terms of frost action and saturation and named it *solifluction*. Von Lozinski (1909), as a result of his studies of weathering in the Carpathian Mountains, recognised the former existence of a peculiar climate, which he called a *periglacial climate*. This climate was influenced by the proximity of the Pleistocene ice sheets. It existed in the regions peripheral to the continental ice and was characterised by intense frost action. Cairnes (1912(a), 1912(b)) recognised a peculiar process of down-wastage active in the mountains of Alaska. This process, which he called *equiplanation*, resembles in some respects the peneplanation by weathering of de Terra (1940). Similar phenomena were studied elsewhere in Alaska by Eakin (1914, 1916) and interpreted to be the result of a special kind of solifluction process which he called *altiplanation*. Högbom (1914, 1926) called attention to the spectacular frost phenomena in Spitzbergen and Scandinavia and postulated therefrom a mechanism of *frost-heaving*. The concept of frost action and frost-produced erosion has recently been both unified and enlarged into the concept of *cryoplanation* by Bryan (1946). He considers that, under a frost climate, hill-slopes are reduced by a process of denudation or down-wastage by frost action as described by Cairnes, Eakin and others. The products of frost action are removed by rivers which flow only in the summer season of melting as described by Poser (1936). Cryoplanation is thus comparable and parallel to the peneplanation of temperate regions. From the foregoing it may be concluded that the concept of a peculiar periglacial cycle is not new, but has gradually evolved over a period of fifty years.

The periglacial cycle, in its ideal sense, is illustrated in Fig. 8.3. This series of diagrams shows how a hypothetical hill and valley of a periglacial region might be modified by the continued action of geomorphic processes in the proportions peculiar to this régime. Following the pattern set by Davis (1899) three stages, early, intermediate and penultimate, are described as youth, maturity and old age.

At the outset of the cycle, during the youthful stage (Fig. 8.3(b)),

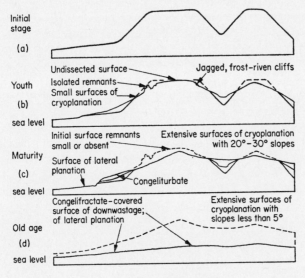

Fig. 8.3   The periglacial cycle

the slopes are attacked by frost-shattering, or congelifraction and are both steep and jagged. Joint patterns must play a strong part in the detailed configuration of such rock slopes. Whenever the joints are widely spaced cliffs may predominate and a talus of coarse angular debris may accumulate at the base; wherever the joints are closely spaced the slopes may be more gentle and the talus may more closely resemble a scree of fine fragments (Behre, 1933) which is the rock-fragment or shale-fragment slope of Judson (1949) and Peltier (1949). This talus may, in the presence of adequate moisture, be subjected to movement and erosion by freezing and thawing as has been described by Capps (1910) and Russell (1933). As the talus grows it provides a lower limit for the attack of weathering upon the exposed rock of the slopes and a continuous slope begins to develop which, in the upper portions, is formed by a congeliturbate mantle lying upon rock, and in the lower portion consists of congeliturbate lying over and incorporating the talus. The bevelled, congeliturbate-covered rock surfaces in the upper part of these slopes are the early phases in the development of a surface of cryoplanation. These slopes may have gradients of about 15 to 20 degrees. At first they are neither wide nor long. They are separated, one from the other, by steep-sided rock valleys with slopes of 25 to 30 degrees (or even steeper), such as were described from Spitzbergen and Greenland by Poser (1936). In places distinct, steep-sided, angular or subangular bodies of rock may rise abruptly from the gentle congeliturbate slope as though partially submerged in a sea of frost-produced fragments. Such steep-sided rocks have been illustrated and described by Cairnes (1912(b)) and Eakin (1916). Undissected remnants of earlier cycles or previous geologic events may remain in the more remote parts of the hills. Here congelifraction (defined by Bryan, 1946), without the accompanying horizontal displacement of congeliturbation, may be expected. Its depth of development would be controlled by the depth of the zone of ground-water saturation and by the depth of freezing and thawing and, therefore, by the extent of its exposure. The solid rock surface should, thus, assume a gently rounded profile. The stage of youth in the periglacial cycle should be characterised by

(1) gently rounded, congelifractate-covered, undissected upland remnants;

(2) jagged, frost-riven cliffs at the base of which a talus of angular, frost-shattered fragments has formed;

(3) gently sloping, congeliturbate-mantled, steep-sided, re-entrant valleys in the cliffs whose slope is continuous with the top of the talus;

(4) isolated, steep-sided, frost-riven remnants surrounded by the gently sloping, congelifractate-mantled surface of cryoplanation.

The stage of maturity (Fig. 8.3(c)) is marked by the disappearance of the isolated rock remnants, the frost-riven cliffs and the broad, gently rounded, undissected upland, and by the extension of broad surfaces of cryoplanation. This change is brought about by the retreat of the slopes bordering the small surfaces of cryoplanation developed during youth and as the result of destruction by congelifraction of all exposed rock where water is present and freezing and thawing frequent. The slope retreat is the result of denudation by congeliturbation, soil-flowage and the slopewash effected by thaw waters and rainstorms. Except in those places which are continuously and effectively subject to river or wave erosion, the products of this denudation will accumulate at the foot of the slope. Wherever this accumulation develops and grows it establishes a gradually rising local base-level which limits the denudation. As a result of this process the congeliturbate-mantled slopes, under the periglacial cycle, may be expected to retreat by successively decreasing gradients. The frost-riven cliffs of the youthful stage, on the other hand, probably retain their steepness as they retreat until they are ultimately consumed as the gradually rising slope of cryoplanation is extended. The extension of surfaces of cryoplanation is most actively carried on by congelifraction concentrated at a horizon within the ground where water is abundant and where the fluctuations of soil temperature about the freezing-point are numerous. Thus, wherever the ground-water table occurs within a zone of frequent alternations of freezing and thawing, very severe congelifraction should occur.

In shallow valleys where the water table lies near or at the surface, this may result in the development of a surface accumulation of congelifractate. Beneath the hilltops, where the water table lies at greater depth below the surface, the zone or horizon of most intensive congelifraction may lie beneath the surface, buried by less thoroughly fragmented material. Its magnitude and significance are determined by the effect of thermal insulation of the overlying rock and by the fluctuations of the water table. The congelifractate thus formed may move gradually down the slope under the action of the same processes which produced it. At the point of inflection between steep and gentle slopes, where in warmer climates springs might be expected, this concentrated congelifraction may undermine the slope, cause it to collapse as a slice and lead to its parallel retreat. This process of mechanical ground-water weathering, determined by the contour of the ground-water table, will lead to the progressive lowering of the hilltops and the progressively increasing radius of curvature of the hilltop profiles in periglacial regions. The surfaces of cryoplanation are therefore developed by both weathering (congelifraction) and erosion (congeliturbation).

The broad surfaces of cryoplanation, developed during maturity, are assumed to be smooth and unbroken by sharp or distinct valleys.

Their gradient is relatively steep, varying between 20 to 30 degrees on the upper slopes and less than 5 degrees near the valley bottoms. Because frost action, including congelifraction, acted over a longer period of time, in producing a mature stage, than was required for the youthful stage, the fragments in the congeliturbate should also have been reduced to finer and smaller sizes than had previously prevailed. The mature stage of the periglacial cycle should, then, be characterised by

(1) long, smooth, gently sloping, undissected, congeliturbate-mantled slopes;

(2) broadly rounded hilltops and hillcrests;

(3) broad, gently sloping, congelifractate-covered valleys or broad, flat valleys filled with congeliturbate and coarse alluvium derived therefrom;

(4) the absence of cliffs and remnants of the previous cycle.

The old-age stage of the periglacial cycle (Fig. 8.3(d)) is the product of continued congelifraction and congeliturbation acting upon the gently sloping surfaces and rounded ridges of the mature stage. Two features characterise this stage. The continued mechanical ground-water weathering and the continued downhill movement of congeliturbate have led to the destruction of the hills and the reduction of the slopes to gradients of less than five degrees. The congeliturbate, by this time, must also be thoroughly comminuted by congelifraction. The sand and silt-sized particles produced by this process (Zeuner, 1945) are vulnerable to wind action. Loess and sand deposits on the one hand and wind-swept pebble-pavements on the other may, therefore, be important during this stage, for the relief is now so reduced as not to seriously interfere with the wind, and plant cover may only locally protect the surface.

The dominant agents of planation active in the periglacial cycle, frost action and wind action, are theoretically independent of ultimate base-level. Wind action is relatively so ineffective that the possibility of a complete cycle due to the wind need hardly be considered. The products of frost action are, however, significant and must be removed, or equiplanation as postulated by Cairnes (1912) would be universal. The work of streams in removing the waste of frost action determines the local base-level. The periglacial cycle may, therefore, develop and go to completion either about a through-flowing drainage or about closed basins. However, should ultimate old age be attained, it must be related to sea-level. Such a plain must, then, resemble in kind, if not extent and coarseness of surface material, those of Lehtovaara and Skansberget in northern Sweden described by Högbom (1926).

EVIDENCE OF A PERIGLACIAL CYCLE

Evidence of a periglacial cycle is widespread throughout the subarctic and high mountain regions. To these areas may be added regions in which a periglacial régime prevailed during parts of Pleistocene time. Most of the existing information relates to characteristic deposits, as relatively little attention has been given to the erosion surfaces which are generally concealed by these deposits. The concept of planation by frost action was first mentioned by Wright (1910). He considered that it might occur above the limit of glacial ice and thus explain some topographic features he observed in Iceland. He did not, however, particularise on the process. Cairnes (1912(a), 1912(b)) described a surface of periglacial erosion (cryoplanation) in Alaska as an example of a process called equiplanation. He, however, focused his attention upon the cut-and-fill action of the process rather than upon its frost-influenced nature. The name equiplanation therefore became applicable to any region of interior drainage, even though he described it as characteristic of the arctic. Eakin (1914, 1916) added his observations of the process he called altiplanation. He described it as a special phase of solifluction that, under certain conditions, expresses itself in terrace-like forms and flattened summits and passes that are essentially accumulations of loose rock materials. This concept was applied to the interpretation of the top of Mount Washington, New Hampshire, by Antevs (1932), who argued that frost planation had produced the lawns, benches and spurs of the Presidential Range. It was further amplified by de Terra (1940), who described the process of cryoplanation in detail as applied to the tendency toward peneplanation by weathering of the Tibetan Plateau.

The effect of periglacial processes in the widening of valleys and the retreat of slopes has been observed in the Falkland Islands by Anderson (1907), in Scandinavia by Högbom (1926), and in Spitzbergen and Greenland by Poser (1936). Their studies show that the existence of saturated ground within the zone of freezing and thawing will lead to the development of a horizon of intense congelifraction and the development of a flat, rubble-covered surface. This surface may extend itself laterally by the action of frost at the base of the valley walls and the undermining of those walls. These surfaces are therefore analogous to surfaces of lateral planation of the moderate, or 'normal', cycle as described by D. W. Johnson (1931), and the valleys may be so widened as to assume the appearance of maturely eroded river valleys. However, during the spring 'break-up', the great rivers of the north also cut laterally, much like rivers of temperate lands, and hence accomplish *lateral planation*. At the same time rock fragments torn from their parent rock by frost-shattering may, either by frost-heaving and subsidence or by plastic flowage of the surficial thawed zone over the permanently frozen subsoil, move down the slopes toward the centre of the valley. This process smooths and extends the slopes and decreases their gradients. Its effects are analogous to those of slopewash under a pluviofluvial régime. The twofold nature of peneplanation, comprising denudation and lateral planation, which is apparent in the moderate cycle, is also present in the cryoplanation of the periglacial cycle. It may be comparable to the distinction, made by de la Noë and de Margerie (1888), between slopes produced by rainwash on hillsides and slopes produced by rivers and streams.

POLYGENETIC TOPOGRAPHY

If, as suggested earlier in this discussion, the geographic cycle is to be interpreted in terms of the peculiar climatic régime under which it was formed, the moderate cycle constitutes one of several distinct and equally important régimes. Between these régimes there may occur transition phases and areas of the invasion of first one and then another climatic régime as the climatic zones have shifted or changed their properties. Wherever these changes are so recent that the effects of the previous geographic cycle have not been erased, polygenetic topographies result. These topographies are the result of two or more superimposed morphogenetic régimes.

Such changes have already been recognised in weathering products and their effects described as polygenetic soils (Bryan and Albritton, 1943). They are equally apparent in the development of periglacial river terraces, as described by Soergel (1924), for here the valley-fill is produced under one régime and the erosion of that fill to produce a terrace under a quite different régime. A periglacial terrace may be regarded as the result of the superimposition of the moderate cycle upon the periglacial cycle. Equally impressive is the evidence, found in both northern Missouri and northern Pennsylvania, of the dissection of the glacial deposits and glacially moulded topography by the stream action of the present. Here, there is a superimposition of the moderate cycle upon the glacial cycle. In many instances, particularly in areas unglaciated since early Wisconsin time, the periglacial cycle has been superimposed upon the glacial cycle and has, in turn, been superimposed upon by the moderate cycle.

The concept of polygenetic topography may be applied to the interpretation of slopes. This is possible because a slope which is

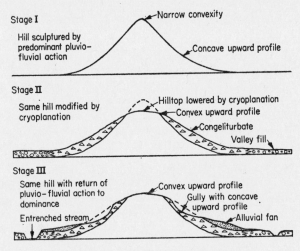

Fig. 8.4 Polygenetic topography

formed, and which is stable under one régime, differs from the slope of another régime and is unstable under the different conditions of climate and plant distribution (see Lowdermilk, 1934). The slopes produced under a periglacial régime, by the dominant activity of frost action, tend to have gentle gradients, to be nearly straight or slightly convex upward in their middle section and to have broadly rounded tops. On the other hand, under a pluviofluvial régime, the action of running water requires that the water be accumulated into a stream to achieve its maximum erosive power. Because the water accumulates with distance down the slope, erosion tends to increase toward the foot of the slope. This tendency is counteracted by the increase in stream load and by the accumulation of alluvium at the foot of the slope. The resultant curve is concave upward. This curve is characteristic of slopes developed under the dominant influence of running water.

In a region such as the state of Missouri, which has been subjected to both régimes, a composite product, such as that shown in Fig. 8.4, must result. Here it is supposed that the slope was modified by the periglacial cycle, from the moderate cycle, which caused an increase in the radius of curvature of the hilltop, straightened, lengthened and smoothed the hillsides, and filled up the valleys. It is then supposed that the slopes became stabilised by the encroachment of a cover of natural vegetation so that this relic topography was retained, even after a change back to the climate of the moderate cycle, and in fact until historic times. Exposure of this slope by deforestation and cultivation, however, has exposed the hills of Missouri to the unimpeded pluviofluvial activity of the moderate cycle. Gullies with a concave-upward profile have developed in the smooth, straight slopes of the periglacial régime. In this instance the contribution of man has been only to tip the scales slightly toward soil erosion and in favour of the existing climatic processes. The latent instability of the slope under the present climate has led to gullying and all the other consequences known as soil erosion.

REFERENCES

ALTER, J. C. (1926). Avalanche at Bingham, Utah. *Mo. Wea. Rev.*, **54**, 60–61

ANDERSON, J. G. (1906). Solifluction, a component of subaerial denudation. *Jour. Geol.*, **14**, 91–112

—— (1907). Contributions to the geology of the Falkland Islands. *Wiss. Erg. d. Schwed. Sudpolar Exped.* 1901–1903, Stockholm, **3**, 2, p. 38

ANTEVS, E. (1932). *Alpine zone of Mount Washington Range*, Auburn, Me., p. 118

ARRHENIUS, S. (1889). Über die Reaktionsgeschwindigkeit bei der Inversion von Rohrzucker durch Sauren. *Zeitsch. f. Physikal. Chem.*, **4**, 2, 226–48

BEHRE, C. H., Jr. (1933). Talus behaviour above timberline in the Rocky Mountains. *Jour. Geol.*, **41**, 622–35

BLACKWELDER, E. (1925). Exfoliation as a phase of rock weathering. *Jour. Geol.*, **33**, 793–806

BLUMENSTOCK, D. I., and THORNTHWAITE, C. W. (1941). Climate and the world pattern, in *Climate and Man*, Yearbook of Agriculture, U.S.D.A., 98–127

BORNHARDT, W. (1900). *Zur Oberflächengestaltung und Geologie Deutsch-Ost-Afrikas*. D. Reimer, Berlin, p. 595

BRYAN, K. (1940). The retreat of slopes. *Ann. Ass. Am. Geographers*, **30**, 254–68

—— (1946). Cryopedology—the study of frozen ground and intensive frost action with suggestions on nomenclature. *Am. Jour. Sci.*, **244**, 622–42

BRYAN, K., and ALBRITTON, C. C. (1943). Soil phenomena as evidences of climatic changes. *Am. Jour. Sci.*, **241**, 469–90

BÜDEL, J. (1937). Eiszeitliche und rezente Verwitterung und Abtragung im ehemals nicht vereisten Teil Mitteleuropas. *Pet. Mitt.*, erg. nr. 229, p. 71

—— (1944). Die morphologischen Wirkungen des Eiszeitklimas im gletscherfreien Gebiet. *Geol. Rundschau*, **34**, 482–519

—— (1948). Die klimamorphologischen Zonen der Polarländer. *Erdkunde*, **2**, 22–53

CAIRNES, D. D. (1912(a)). Some suggested new physiographic terms (equiplanation, deplanation and aplanation). *Am. Jour. Sci.*, **34**, 75–87

—— (1912(b)). Differential erosion and equiplanation in portions of Yukon and Alaska. *Geol. Soc. Am., Bull.*, **23**, 333–48

CAPPS, S. R., Jr. (1910). Rock glaciers in Alaska. *Jour. Geol.*, **18**, 359–75

COTTON, C. A. (1942). *Climatic accidents in landscape-making*. 2nd ed., Whitcombe and Tombs, Christchurch, p. 354

DAVIS, W. M. (1894). Physical geography as a university study. *Jour. Geol.*, **2**, 66–100

—— (1899). The geographical cycle. *Geographical Jour.*, **14**, 481–504

—— (1900(a)). Glacial erosion in France, Switzerland and Norway. *Proc. Boston Soc. Nat. Hist.*, **29**, 273–322

—— (1900(b)). Physical geography in the high school. *School Rev.*, **9**, 388–404

—— (1909). Complications of the geographical cycle. *Proc. 8th Internat. Geog. Cong.*, 150–63 (see also *Geog. Essays*, ed. by D. W. Johnson, Ginn & Co., 1909)

DE TERRA, H. (1940). Some critical remarks concerning W. Penck's theory of piedmont benchlands in mobile mountain belts. *Ann. Ass. Am. Geog.*, **30**, 241–47

EAKIN, H. M. (1914). The conditions of 'altiplanation' in subarctic regions (abstr.). *Jour. Wash. Acad. Sci.*, **4**, p. 171

—— (1916). The Yukon-Koyukuk region, Alaska. *U.S.G.S., Bull.* No 631, p. 88

FREISE, F. W. (1938). Inselberge und Inselberg-Landschaften im Granit- und Gneisgebiete Brasiliens. *Z. f. Geomorph.*, **10**, 137–68

GLASSTONE, S. (1940). *Textbook of Physical Chemistry*. Van Nostrand, New York, p. 1289

HETTNER, A. (1921). *Die Oberflächenformen des Festlandes, ihre Untersuchung und Darstellung*. Teubner, Leipzig, p. 250

HÖGBOM, B. (1908–9). Einige Illustrationen zu den geologischen Wirkungen des Frostes auf Spitzbergen. *Bull. Geol. Inst. Upsala*, **9**, 41–59

—— (1914). Über die geologische Bedeutung des Frostes. *Bull. Geol. Inst. Uppsala*, **12**, 257–389

—— (1926). Beobachtungen aus Nordschweden über den frost als geologischer faktor. *Bull. Geol. Inst. Uppsala*, **2**, 243–79

HOOTS, H. W. (1930). Geology and oil reserves along the southern border of San Joaquin Valley, California. *U.S.G.S., Bull.* 812-D, 243–332

JOHNSON, D. W. (1919). *Shore Processes and Shoreline Development*. Wiley, New York, p. 584

—— (1931). Planes of lateral corrasion. *Science*, **73**, 174–77

JUDSON, S. (1949). Rock-fragment slopes caused by past frost action in the Jura Mountains (Ain), France. *Jour. Geol.*, **57**, 137–42

KESSLER, P. (1925). *Das eiszeitliche Klima und seine geologischen Wirkungen im nicht vereisten Gebiet*. Stuttgart, p. 210

—— (1927). Über diluviale Frostspalten bei Saarbrucken. *Z. d. Deutsch. Geol. Gesel.*, **79**, 75–80

KÖPPEN, W. (1923). *Die Klima der Erde.* De Gruyter, Berlin, p. 369

LOWDERMILK, W. C. (1934). Acceleration of erosion above geologic norms. *Trans. Am. Geophys. Union,* Pt. 2, 505–9

LOZINSKI, W. VON (1909). Über die mechanische Verwitterung der Sandsteine im gemässigten Klima. *Bull. Acad. Sci. Cracovie,* 1–25

MATTHES, F. E. (1900). Glacial sculpture in the Bighorn Mountains, Wyoming. *U.S.G.S., Ann. Rpt.* no. 21, pt. 2, 167–90

NOË, G. DE LA, and MARGERIE, E. DE (1888). *Les formes du terrain.* Sery. Géog. de l'Armée, Paris, p. 205

PASSARGE, S. (1926). Geomorphologie der Klimazonen oder Geomorphologie der Landschaftsgürtel? *Pet. Mitt.,* 72nd Jhg., 173–75

PELTIER, L. C. (1949). Pleistocene terraces of the Susquehanna River, Pennsylvania. *Penn. Topog. and Geol. Surv., Bull.* G-23, p. 158

PENCK, A. (1905). Climatic features in the land surface. *Am. Jour. Sci.,* 4th ser., **19,** 165–74

—— (1910). Versuch einer Klimaklassifikation auf physiographischer Grundlage. *Sitz.-Ber. Preuss. Akad. d. Wiss., Phys.-Math.* Kl., **12,** 236–46

POSER, H. (1936). Talstudien aus Westspitzbergen und Östgrönland. *Z. f. Gletscherkunde,* **24,** 43–98

REICHE, P. (1937). The toreva-block—a distinctive landslide type. *Jour. Geol.,* **45,** 538–48

RUSSELL, R. J. (1933). Alpine landforms of western United States. *Geol. Soc. Am., Bull.,* **44,** 927–50

SAPPER, K. (1935). Geomorphologie der feuchten Tropen. *Geogr. Schriften,* **7,** p. 154

SHARPE, C. F. S. (1938). *Landslides and related phenomena.* Columbia University Press, p. 137

SMITH, H. T. U. (1936). Periglacial landslide topography of Canjilon Divide, Rio Arriba County, New Mexico. *Jour. Geol.,* **44,** 836–60

SOERGEL, W. (1924). *Die diluvialen Terrasen der Ilm und ihre Bedutung für die Gliederung des Eiszeitalters.* Fischer, Jena, p. 76

STECHE, H. (1933). Beiträge zur Frage der Strukturboden. *Ber. üb. d. Verhandl. d. Sächs. Akad. d. Wiss.,* **85,** 193–250

THORBECKE, F. (1927). ed. Morphologie der Klimazonen. *Düsseldorfer Geogr. Vorträge u. Erörterungen,* p. 99

TROLL, C. (1944). Strukturboden, Solifluktion und Frostklimate der Erde. *Geol. Rundschau,* **34,** 545–694

—— (1947). Die Formen der Solifluktion und die periglaziale Bodenabtragung. *Erdkunde,* **1,** 162–75

—— (1948). Der subnivale oder periglaziale Zyklus der Denudation. *Erdkunde,* **2,** 1–21

WENTWORTH, C. K. (1943). Soil avalanches on Oahu, Hawaii. *Bull. Geol. Soc. Am.,* **54,** 53–64

WRIGHT, F. E. (1910). Some effects of glacial action in Iceland. *Bull. Geol. Soc. Am.,* **21,** 717–30

ZEUNER, F. E. (1945). *The pleistocene period: its climate, chronology and faunal successions.* Ray Soc. London, p. 322

9 The Problem of Erosion Surfaces, Cycles of Erosion and Climatic Geomorphology

HERBERT LOUIS

W. M. DAVIS (1899), with his extension of G. K. Gilbert's notion of the geographical cycle and its end product the peneplain, has given the study of landforms a fascinating and stimulating basic idea. It is true that objections have been raised, particularly in Germany, against Davis's views. A. Hettner (1921) and S. Passarge (1912), in particular, have pointed out several uncertainties in Davis's deduction and stress the necessity for a more thorough examination of climatic influence on the exogenic processes and the vagueness of young, mature and old as a terminology to describe characteristics of landforms. Both A. Penck (1919) and W. Penck (1924) have examined the possibility of the simultaneous interaction of crustal movements and exogenic processes, something which was not sufficiently explored in Davis's work. The difference between *Endrumpf* and *Primärrumpf*, the idea that flat as well as gently sloping and steep landforms could be the expression of a balance between uplift and depression, and the subsequent view that, with the passage of time, not only a succession of steep, gentle and flat forms is possible, but also any desired succession of these, was a result which went far beyond the framework of Davis.

In essence, however, all these ideas are just refinements of Davis's basic tenet, namely that the landforms of any piece of the earth's surface are the expression of tectonic movements in the crust and of the exogenic processes which affect it.

Meanwhile, research has produced important new information. The most significant is that, apart from the long-recognised differences between fluvial, glacial and arid landforms, there exist important climatically caused differences within areas where river action is all-important. A new discipline of climatic geomorphology is coming into being. But one thing that so far has been little examined is the question of what influence these new ideas will have on the former basic assumptions. We would maintain that, in fact, they deeply affect the fundamentals of Davis's ideas and force a new conflict with Davis's basic doctrine of the geographical cycle and peneplain (Penck's *Endrumpf*).

DAVIS'S BASIC HYPOTHESIS

The assumption that in the course of erosion, if no processes interrupt or divert the development then a plain-like surface must be formed very close to sea-level representing the final product of the erosion process, is fundamental to Davis's model. From the purely logical point of view this idea may be viewed as both unquestionable and convincing. Davis himself, however, has demonstrated that the final stages of this process require an exceptionally long period of time, since the gradient of the rivers becomes increasingly less. The question has already been asked as to whether such long stable periods have ever occurred in the earth's history. Davis decided that they must have happened because of the actual occurrence of the erosion surfaces. He noted further that in many mountainous regions such erosion surfaces existed, and concluded that periods of uplift had raised these flat landform features which had been formed at low altitudes.

Davis recognised that in arid areas the base-level of erosion can lie high above sea-level and that, theoretically, a peneplain could be formed locally at such a high level. Since plant remains and even coal have been contained in strata at this height in conjunction with these erosion surfaces, that is, evidence of a humid climate with discharge to the sea, he therefore concluded that the special case of aridity with a high base-level involves no basic objection to the general theory of the peneplain or *Endrumpf* surface close to sea-level.

It is really fundamental to Davisian thought that the *Endrumpf* surface can only develop at a height just above sea-level as the ultimate base-level. Otherwise a major application of this theory would disappear, namely the view that the occurrence of such elevated and dissected erosion surfaces can with certainty point to relative uplift of the land after the formation of the erosion surface.

THE FORMATION OF EROSION SURFACES AT CONSIDERABLE HEIGHTS IN TROPICAL REGIONS WITH WELL-MARKED SEASONALITY

Research over the past 50 years has clearly shown that the present formation of erosion surface-like 'flat forms' not only occurs close to sea-level, but often occurs at considerable heights in landscapes that drain to the sea. Bornhardt (1900) in his classic description of the 300–400 m high gneiss surfaces in Madjedje in the hinterland of Lindi in East Africa illustrates this. Here too are the prototypes of the inselbergs that rise so steeply and sharply from the surrounding country. These surfaces of low relief will gradually be extended with backward erosion of the inselbergs. Thus an increasing levelling of the earth's surface is taking place at a considerable height above sea-level.

N. Krebs (1933) has reported an 80 km-wide rock plain in Madura and Tinevelli (India) which rises to 200 m with a two to three per cent gradient, and which exhibits inselbergs. According to O. Jessen (1936) the surfaces of the Planalto in Angola (expressionless highland surfaces with broad flat-floored valleys (*Muldentäler*) and with rolling interfluves, which cut right across Archaean gneisses and schists, Algonkian metamorphics, and Algonic and Paleozoic intrusives) reach heights of over 1700 m. Their lateral extent is also increasing, as erosion and weathering gradually wear away the encircling highland (for example *Bimbe*). Thus the formation of erosion surfaces (*Rumpfflächen*) in areas which have outlets to the sea is at present taking place at altitudes well above 1000 m. Of course, these features are by no means common to all areas where running water is the main agent of erosion and transport; they occur only in certain climatic types, above all in savanna regions, or tropical areas with alternating wet and dry seasons.

THE DESTRUCTION OF EROSION SURFACES IN MIDDLE LATITUDES

In many other climates there is no sign of present-day erosion surface formation, but only evidence of the gradual destruction of former surfaces. The forest zone of the humid middle-latitudes belongs to this category. The very low-lying erosion surfaces which are found by the coast, like those in western France which stretch from Normandy to the Gironde, are especially instructive. From Cotentin to the Vendée we are dealing with the erosion surface of the Armorican Massif, which is astonishingly low close to the sea. Some areas should be especially mentioned: de Lessay in western Cotentin, the coastal fringe at Plouescat north-west of Brest, or that of Ploudalmézeau north of the town, the coastal fringe south-west of Quimper or the low-lying coastal area of the Morbihan and Vilaine. All these low-erosion surfaces are being deeply incised by streams (*Kerbtaleinschritten*); they are no longer being extended but rather are slowly being destroyed and, in fact, this has been in progress for a fairly long time. These low but dissected erosional landscapes are not confined to the Armorican Massif; the same conditions are to be found in the Campagne of Caen on the lower Orne where the low flat landscape extends over Jurassic strata, and also in the area around La Rochelle.

A similar situation occurs on the east coast of North America in those areas which were not affected by the former inland ice cover, that is south of the Hudson estuary. The erosion surface, which here extends over the ancient folds of the piedmont, lies along the line Philadelphia–Baltimore–Washington–Richmond at only 60 to 80 m above sea-level at distances of at least 100 km from the sea. This surface is also deeply dissected by steep-sided valleys, like the low-lying deposits closer to the coast which cover the older rocks of the coastal platform. Quite clearly, these extensive erosion surfaces of the piedmont and the surrounding coastal lowland, despite their low altitude above sea-level, are at present not being extended but, as former plains, are being destroyed.

A similar situation has also been observed in East Korea. According to Kobayashi, an extensive low-lying erosion area with monadnocks has developed in the areas around Seoul and Pjongjang, the so-called *Rakuro*. This erosion surface lies at 50 m above sea-level 25 km from the coast, and then rises to 200 m about 70 km inland. Present-day rivers with broad channels and steep sides have cut into this surface too, so that it is not being enlarged; on the contrary present processes are working towards its slow destruction.

If we are to explain the origin of these steep-sided valleys by a relative uplift of the land after the completion of the final erosion surface (*Endrumpf*), then the amount of uplift or tilting required to achieve this enormous downcutting would appear to be minimal. A movement of less than 20 m would have been necessary in the French coastal areas and certainly no more than 50 m in other areas. It may therefore be assumed that a small change of sea-level would greatly interrupt the process of erosion–surface formation. If one thinks, however, of how replete geological time is with transgressions and regressions—not to mention the eustatic changes of the Quaternary—then it becomes very difficult to believe in the long periods of absolute crustal stability and constancy of sea-level which, according to Davis, account for erosion surfaces near sea-level.

This difficulty does not arise in climatic–geomorphological theory. It sees these erosion surfaces as Tertiary erosion surfaces, which were formed regardless of height above the sea. There are remains of former weathered soils on these surfaces, which are similar to the soils of present-day savanna areas. They show that gently rolling landscapes (*Flächmuldenlandschaften*) can in fact be formed at these altitudes. Present-day dissection shows it to be a geomorphological expression of the climatic changes that have occurred since its formation. It points to the fact that these highland areas (for example the surfaces of Madura and Tinevelli, the high plains of Canvery and also those of Angola), would have been immediately dissected if transferred to our humid middle-latitude climate.

In order to make a comparison between the geomorphological individuality of these elevated *Rumpf* landscapes of the savannas and middle-latitude landscapes, one would have to think of a Europe in which neither the Erzgebirge, the Frankenwald, the Rhine Highlands, the Ardennes, Brittany nor the French *Massif Central* had been dissected by streams. All these highland areas rise on the whole so gently that, apart from isolated steeper scarps, even the larger rivers would flow in gently hollowed valleys over these rises, if a savanna type of climate prevailed. As far as the 'steps' are concerned they would certainly be cut by gorges, but these would only partly intrude into the area of the erosion surface above the 'step'. This is the case as shown by research into stepped erosion surfaces (*Rumpftreppen*) in the seasonally wet tropics.

From all this it may be concluded that these erosion surfaces are not in fact *Endrumpf* surfaces, nor have they ever been such, but rather they are features which have been formed under certain climatic conditions on a massive, undissected relief basement (*Reliefsockel*), that is they do not approach the theoretical ultimate stage of terrestrial erosion and transportation.

Neither are these erosion surfaces *Primärrumpfflächen* in the sense of Walther Penck for according to Penck's definition this should be a flat erosion surface, formed during gentle uplift, by erosion which keeps pace with the uplift. The question then arises as to how such erosion surfaces, which do not approach sea-level and which represent neither *Endrumpf* nor *Primärrumpf*, are formed.

THE DISTINCTION BETWEEN FLUVIAL LANDSCAPES OF
INCISED (*KERBTAL*) AND GENTLY-ROUNDED (*FLACHMULDEN*)
TYPE

Since the theories of *Endrumpf* and *Primärrumpf* obviously do not fit the observed facts, any attempt to understand the nature of these erosion surfaces must depend on a theoretical understanding of the interactions of the various process–complexes which are at work in forming them, their modification as they are progressively destroyed, and the formation of other landscape features. This attempt will require the application of abstraction and deduction without straying from the findings of field research.

Our theoretical consideration recognises the importance of the relationship between the surface denudation potential and the rate of linear erosion. The following assumption can be made: because of variability in rock hardness and the vast differences in the amount of weathered material produced under various climates, there must be wide spatial variation in denudation potential and a variable balance between surface denudation and linear erosion.

Two extreme cases may be cited. In the first instance, denudation potential is low, and linear erosion creates such high and steep-sided valley slopes that denudation capacity is increased sufficiently to load the stream to such an extent that its mechanical energy is used up; thus, during downcutting, the stream keeps pace with the denudation of the slopes.

One result of this process is perhaps best termed a V-shaped valley or *Kerbtal* type of fluvial relief. For our own climatic zone, it is the normal type of valley landscape. The steepness of the slopes varies greatly depending on the nature of the rock. However, it is important that the channels form considerable valleys *before* increased slope denudation provides such an amount of debris, so that the energy required to deal regularly with the work of transportation is in equilibrium with the volume of waste supplied.

The other extreme case concerning the balance between surface denudation and linear erosion occurs with an increased capacity for weathering (*Gesteinsaufbereitung*) and general denudation. The result is a flat erosion landscape with rounded valleys (*Muldentäler*) having a very gentle cross-section. When weathering and associated denudation processes are powerful, then steep-sided slopes are quickly worn down. The channels naturally follow depressions, but they cannot cut into the surface too greatly, because they acquire such a mass of denudation material that their mechanical energy is used up in transporting it. Each incision into the gentle slopes immediately creates adjacent areas of strongly increased denudation with a consequent increase in the amount of material over and above that which is transported. This can be termed the *Flachmulden* (gently moulded) type of fluvial landscape. Its creation is very much bound up with powerful weathering processes, which are climatically induced. It also seems to be favoured by periodic flooding of streams. There does not appear to be any consistency in the heights of such surfaces above sea-level. Such landscapes are generally characteristic of the savanna areas with well-marked dry and wet seasons. There is only one feature which diverges from this description, namely the inselberg. Under these climatic conditions, when angles of slope exceed a certain steepness percolation decreases and there is a decrease in weathering which tends to perpetuate the inselbergs (Jessen, 1936).

Despite the flat nature of the ground, erosion in the *Flachmulden* landscape is by no means inconsiderable. Even the larger rivers, although they are not incised, usually have a considerably greater longitudinal gradient than comparable stretches in a *Kerbtal* landscape. For example, the larger rivers of Central and Western Europe have in the lowermost 200 km of their courses gradients of less than $0.2‰$, while some rivers of the Deccan have gradients of $0.4‰$ in their lower reaches.

From this we can say that in *Flachmulden* landscapes more energy is used up in the main channel per unit of distance than in corresponding channels of *Kerbtal* landscapes. This increase can only be explained by an increased transportation of material. Moreover, it must not be concluded that because the rivers are not incised there is no vertical erosion, for it has often been observed that such rivers expose and attack the solid underlying rock. As a rule, of course, the overburden is thin. It appears that subaerial denudation on the extensive gentle slopes in these 'flat' landscapes just about keeps pace with the vertical erosion of the river, so that all the valley slopes remain exceptionally gentle. The lack of coarse debris in these rivers, a consequence of climate, naturally lessens the downcutting capacity of the channels.

Accordingly, it is evident that the *Flachmulden* landscape, and the processes involved in forming it are completely different from those conceived by Davis as those responsible for the production of the peneplain surface for, according to Davis, the volume of erosional material should become very small with decreasing height above sea-level. Neither is the *Flachmulden* type of landscape a *Primärrumpf* in the sense of W. Penck. For, according to the definition, a *Primärrumpf* can only be formed at very low altitudes. Furthermore, it should exhibit a convex-slope profile, which Penck claims to be character-istic of his waxing-slope sequence (*aufsteigenden Entwicklung*). On the contrary, the *Flachmulden* landscape exhibits a concave-slope profile, which, according to Penck, is a criterion for his waning-slope sequence (*absteigenden Entwicklung*). This explanation hardly seems to suffice for those high altitude *Flachmulden* landscapes of the savanna regions. It certainly appears that great care must be exer-cised if slope profiles are to be related to crustal movement.

A comparison of the natural environment shows that both these extreme forms, that is the *Kerbtal* and *Flachmulden* types of land-scape can, under certain climatic conditions, occur in a particularly pure form. But it is to be expected that Nature will exhibit few such extreme examples of the interaction of surface denudation and linear transport. Because of the multiplicity of variables involved, we can further assume that between these two extremes there is no simple progression from one type to another, but numerous variations of linked landforms. Some indication of this can be gleaned from an examination of the landform characteristics of a climatically 'one-sided' area.

LEVELLING PROCESSES IN THE POLAR SOLIFLUCTION ZONE

The solifluction zone, made up of areas either near the Pole or above the treeline, belongs to those fluvial erosion landscapes which are strongly influenced by climate. In this zone, frost weathering and solifluction are the major denudation factors. Büdel (1948), Poser (1932, 1936) and Troll (1947, 1948) have all contributed to this field of study in Germany. Powerful, seasonal meltwater torrents are responsible for transporting the load. In these areas there obviously exists very intensive interaction between surface denudation and linear transportation which quickly begins to create new landforms. Büdel has given a host of examples of Quaternary cirques and troughs in the frost-shattering zone which have been largely obliter-ated by such processes. The so-called macrosolifluction, or marked downhill soil movement, occurs on slopes as gentle as 2°. Nevertheless, these landscapes may be dissected by *Kerbtäler*. To become overloaded, the meltwater torrents must increase the denudational slopes by downcutting. (Unfortunately there is as yet little information on the gradients of the large, broad valleys, whose gravel beds are not usually connected by terraces.) From this, and with due regard to the intensity of denudation, conclusions can be drawn as to the total erosion potential of these areas which, as Büdel has stressed, appears to be very great. Unfortunately, flat erosional landscapes in rocky areas in this fairly recently deglaciated climatic zone have not been as thoroughly investigated as the mountain landscapes of Spitzbergen and East Greenland. One may assume that in such landscapes macrosolifluction as a denudation process must finally overrule everything, and that because of the tendency of the heavily debris-laden meltwater streams to undercut the valley slopes, a landscape must be formed with a flat slope profile, slightly convex at its lower edge, and with broad valley bottoms. Thus, one might speak of a *solifluctionsrumpf* landscape.

This, then, is another erosion surface of climatic–geomorphological type which apparently can be relatively quickly created. There is no implication that such a surface in any way approximates the end product of terrestrial erosion in the sense of Davis's peneplain. For example, with a 2° slope, relative relief of up to 100 m is possible over a distance of two-and-a-half km. Such solifluction erosion surfaces may well occur in the flat highlands above the treeline, for example in Tibet and the Andes, as Troll has already stated.

The character of such a solifluction surface differs from that of the *Flachmulden* landscape of the savanna not only in its tendency towards *Kerbtal* downcutting, but also in the character of its higher relief remnants. Isolated individual hills retain gentle slopes thanks to macrosolifluction, and not the steep sides and abrupt basal change in slope shown by tropical inselbergs. This must be taken into account in the case of many old erosion surfaces in the middle latitudes, since they have been affected not only by Tertiary and present-day erosional forces but also by those typical of the tundra climate which prevailed in the Quaternary.

EROSION SURFACES IN ARID AREAS

Together with the savanna and the frost-weathering zones, arid regions have long been recognised as areas where erosion surfaces of low relief are formed. It has been observed that the flat floors of the enclosed inland basins, like the bolsons of the western United States, do not consist only of deposition material, but have rocky erosion surfaces under a thin layer of debris; and above these, those parts of the mountains that have not yet been worn down rise abruptly with a clearly distinguishable change of slope.

According to von Wissman (1951), erosion surfaces in arid areas are formed by the lateral erosion of debris-laden channels at the *foot* of mountain areas, which themselves have been dissected by valleys of the incised or *Kerb* type. The slopes of these mountain areas obviously provide a greater quantity of debris than slopes in middle latitudes, which are covered with vegetation. This accounts for the amount of debris carried by the streams and their relatively steep profiles. Consequently, the erosion surface which develops at the foot of the mountains is very considerable. The slopes increase from the interior of the bolson to about 5°, in places even up to 10°, at the mountain foot where it is deeply dissected by stream valleys (*Tiefen-kerbung* as used by von Wissman). Level surfaces are formed, then, at the altitude of the inland drainage basin which, in many cases, may be well above sea-level. It represents a completely different pheno-menon from Davis's peneplain surface, even if one disregards the height above sea-level. Even if, as sometimes occurs, all irregularities above the erosion surface are removed, that is if a *panfan* (a name coined by Lawson, 1915) is formed, then it has nothing whatsoever to do with Davis's peneplain, but with an erosion surface formed under special climatic conditions and having a considerable slope over a huge altitudinal range. This is still a long way from Davis's ultimate product of terrestrial erosion, even if one were to take into consideration only the elimination of the relative relief down to the level of the bolson floor.

The pediment surfaces of arid areas are, however, something basically different from the rolling *Flachmulden* landscapes of the savanna zone. Unlike these they are not the product, even in gently sloping areas, of an equilibrium between surface denudation and downcutting of the larger channels: rather, they have affinities with the *Kerbtal* type of landscape. In fact *Kerbtal* dissection is typical of arid areas. Although mechanical weathering is extensive, there is considerable surface denudation because of the lack of percolation into the cohesive surface of the broad slopes. Therefore, a periodic water course may cut down almost unhindered. It forms rain gullies

(*Racheln*) and badland-type landscapes. But these incisions cannot become as deep as their equivalents in humid regions for, with down-cutting, the slopes gradually become steeper. This mobilises a considerable amount of debris and the stream becomes so laden that it requires a relatively steep gradient to transport the debris, thus preventing undue deepening of the valley. Thus valley depths depend on the water discharge and the frequency of high floods in the channels, quite apart from the speed of weathering of the rocks themselves. As chemical weathering in this dry climate is not very effective, the channels are provided with abundant coarse material. This coarse bedload constitutes the tool with which extensive rock-cut pediment surfaces are produced by the process of lateral stream erosion below the point where valleys emerge from the mountains.

In process and form the pediment surfaces of the arid zone differ fundamentally from both the *Flachmulden* landscapes of the savanna and the solifluction surfaces of polar and montane regions. Of course, only a small selection of the climatically extremely varied suite of fluvial landscapes can be mentioned here but it is hoped that it is sufficient to allow us to characterise the nature of the interaction between the various erosion processes. The following general statement may be made: climatic–geomorphological research has shown that under certain climatic conditions (in savanna, arid, polar and mountain regions) extensive level areas have been created relatively quickly by erosion and that these surfaces are formed at a considerable height above sea-level on oldlands. However, other climatic areas, especially the temperate forest zone, are not at all favourable to the formation of such erosion surfaces. Moreover, it is thought that the considerably long period of crustal inactivity and constancy of sea-level necessary for the formation of Davis's peneplain just above sea-level, could never really have occurred. Therefore, it is far more logical to conclude that all erosion surfaces have in fact climatic–morphological origins and that genuine peneplain or *Endrumpf* surfaces in the sense of Davis and W. Penck (erosion surfaces with an extremely low absolute relief) do not form or, at most, constitute an extremely rough approximation to these features.

THE INADEQUACY OF THE CRUSTAL MOVEMENT THEORY OF FLUVIAL EROSION

We have seen that in Davis's cyclic theory and Penck's refinements including the so-called morphotectonics, assertions have been made which are not acceptable. Büdel (1948) has already recognised that the extreme variability of the denudation processes in different climates may have a considerable influence on the character of the fluvial landscape. It is useful to consider where the purely theoretical weaknesses of the old denudation theories lie.

Davis's theory proceeds from the following basic premise: the magnitude of general surface denudation in any area is basically dependent on the amount of linear erosion by water channels, apart from the factor of the nature of the rock and the climatic influence which can either prevent or promote denudation. It is these water channels which, by their downcutting, determine the slopes on which denudation works. Davis's ideas of young, mature and old landforms rest on the acceptance of this premise, as do A. Penck's refinements including his 'perennial surfaces' (*Dauerrumpf*) and his later ideas on linking crustal movement with flat, gentle and steep forms in an erosion cycle. The latter, in their turn, led to W. Penck's theories of waxing and waning slope development and their refinement by Spreitzer (1932). All these workers saw crustal movement as the most important controlling factor in fluvial landscape morphology.

Amongst these theoretical assertions, there is at least one idea that does not possess general validity. It is the theory that the gradient of a stream is determined by the crustal movements which form the relief. This theory neglects the fact that a stream, which has a load of corrasive material, requires more energy to remove this material than a similarly sized channel with less load. The stream can only gain this work capacity from the potential energy released when flowing. Therefore, a heavily laden stream will possess a steeper gradient than a stream of similar power but smaller load. Development is always striving to an approximate equilibrium, by which the energy of the stream is so measured that the gradient provides just enough energy to transport the load, or more exactly a little more, since concurrent with the gradual erosion of the slopes the bed of the stream itself must slowly become deeper.

Under such circumstances, a stream whose slopes provide it with debris can cut less deeply into the surface than an equally powerful stream which receives only a small amount of debris. For this latter type requires less of a gradient to deal with its load. The stream, however, possesses a gentler gradient when it is deeply incised, that is when it has moved closer to its base-level of erosion. On the other hand, the increase in downcutting is at the same time a means by which the channel itself increases the amount of its debris load. Clearly, the higher and often steeper valleyside slopes above a downcutting stream provide more slope debris than the gentler slopes of a slight incision. By downcutting, therefore, the stream can produce a situation in which gradient progressively decreases and load increases so that an approximate equilibrium between debris load and transportation capacity is attained.

Given this, it is obvious that increases or decreases in the rate of subaerial denudation by amounts which could well be climatically conditioned, must play a considerable part in the formation of the gradient of these channels. The relief-forming crustal movements, on the other hand, marginally influence the gradient of the channel only when they create a variable relative relief by differential uplift. Between the highland and the sea, there is an unlimited number of possible curves of similar gradient which flowing water can assume. Which of these curves at any given time any observed stream adopts will depend only in part (and in certain cases only to a very small degree) on the extent and character of the crustal movements associated with the uplifts. For this reason Davis's deductive erosion cycle cannot have general validity, even if theoretically necessary, but in practice highly improbable, requirement of absolute crustal stability during the whole course of the cycle were allowed. It is true that Davis's notion of the succession of 'youthful' and 'mature' forms in the case of the *Kerbtal* type of fluvial landscape may approach reality. But his 'old age' or 'senile' forms are clearly quite distinct on climatic–geomorphological grounds from the *Flachmulden* type of landscape.

The cyclic theory can be applied to the *Flachmulden* landscape of the savanna areas only after considerable rethinking. In it, the stages of 'youth' and 'maturity' are, as far as we know, not time-dependent but spatially variable features associated with low-relief surfaces. They occur in the area of the basal scarps (*Rumpfstufe*), often several hundred metres high, which separate various high erosion surfaces from one another. The level surfaces (*Rumpfebenen*) themselves could, however, be characterised as *apparently* old, for the most important characteristic of genuine age in the Davisian sense (as opposed to apparent age) is lacking here, that is the formation on an oldland of a peneplain near to sea-level.

If, however, the term 'cycle' is retained in English and French geomorphological literature and is used in terms such as normal–humid, polar–humid, arid and in future perhaps even seasonal–tropical cycle of fluvial erosion, then we must make it very clear that the word 'cycle' is no longer used in the all-embracing sense in which it was coined by Davis. It can only be used in the sense of 'course of formation towards a specifically attainable geomorphological goal' and not, as Davis would have it, 'a regulative basic idea of fluvial erosion' with all its inherent consequences.

THE LAW THAT GRADIENT INCREASES UPSTREAM

This examination leads us to some explanatory conclusions for the so-called law of headward erosion, or as it has also been called, especially since J. Sölch (1935), the law of upstream increase in gradient.

The general theory is this: an increase in the regional gradient, perhaps caused by uplift of part of the earth's crust, instigates an increased downward erosion in the river valleys of an area. A given channel will incise more powerfully than before. In consequence, an increased gradient is produced in the valley section upstream of the mouth. This causes an upstream transmission of valley deepening, and so on, resulting in a general decrease in altitude and in the severity of the gradient. One might add the corollary that this type of increase in gradient has the potential to cut back to the source of any given river, even if the process is attentuated with time. This idea is of critical importance for the crustal movement theory of fluvial relief, for from it springs the idea that crustal uplift can influence the character of a valley right into its upper reaches. As we shall now see, however, this assumption is unjustified.

The facts seem to be as follows: due to the postulated uplift, the flowing mass of water above and in the extensive transition zone between raised and non-raised areas acquires a quite marked increase in potential energy. This is transformed into kinetic energy by the flowing water and is practically completely expended by attrition as it flows down the newly created steep gradients, that is, it is transformed by heat generation, transportation of load and the additional attack on the bed of the channel. This is confirmed by the fact that the average speed of through-flow of a river in the stretches below the change of gradient is nowhere very different from that above the change. The attack on the channel-bed causes vertical erosion and this is transferred upstream. So far the concept seems to be completely in order.

However, it is necessary to take into consideration the effect on the valleyside slopes of the previously mentioned retrogressive deepening. As the valley deepens, its slopes get longer. As far as we know certain *Kerbtal* characteristics persist in humid, extratropical landscapes, as well as in both arid and polar areas, if the relative altitudes of such landscapes are very low. Thus the areas which are exposed to denudation, and so provide the river with more material, are gradually extended. Part of the potential energy gained by uplift is now used up in the transportation of the increased load so derived, so that only a portion of it is available for the continuation of downward erosion. This can also be expressed as follows: the increase in energy required for the transport of the increased load is obtainable only in those reaches where the gradient has been increased by headward erosion. This means that the gradient must decline going upstream. From the additional amount of potential energy released by the whole process, an ever-increasing proportion must therefore be expended on the transportation of the increased load. At some point or other in the long profile of the channel there must lie a point at which the potential energy produced by headward erosion is completely used up in transportation due to the downstream increase in load. At this point the 'effective height' of the nickpoint is reduced to nil, and any further upstream migration ceases. The nickpoint, which to start with is a relatively short feature, is completely replaced by a long zone of greatly increased valley gradient. This finds itself in a new equilibrium with the amount of debris delivered by the slopes, which are now higher and thus more capable of being denuded. From this time on, uplift no longer exerts such a specific effect on the character of the relief, although the total work (*Gesamtarbeit*) of the system is increased by uplift. Accordingly, any features that develop with further erosion are completely determined by climatic factors.

How high the nickpoint can migrate upstream depends on three main factors. Firstly, the amount of water in the channel is significant. Since large rivers have a lesser gradient than smaller ones (all other circumstances being equal), then spatial gradient change in a large river is greater than in a small stream. Secondly, it depends on the actual amount of uplift, which in certain cases causes headward erosion. In our view, a large nickpoint can obviously cut back more vigorously than a small one, since it takes longer for its effective height to be reduced to nil. Thirdly, the effectiveness of headward erosion depends primarily on the surface denudation rate of an area. Where the general denudation processes are weak, renewed down-cutting can considerably alter the amount of debris supplied to the stream. The nickpoint can migrate upstream without losing very much of its effective height in the process. Thus it is able to migrate a long way upstream before becoming obliterated. Where the general denudation processes are powerful, however, the opposite occurs. The migrating nickpoint quickly loses effective height and is obliterated at a relatively short distance from its point of origin. The upstream migration of nickpoints is therefore dependent to a considerable extent on climatic conditions.

The *Flachmulden* landscapes of the savanna zones exhibit particularly impressive features of this type. In the planation surface landscapes, the uniform 'steps', several hundred metres in height, are not sufficient to cause even the largest rivers to cut down deeply into the flat surface of the next step above. Only in the vicinity of the step itself and in the reach immediately above it are these rivers incised. The total energy provided by these relatively short stretches is obviously used up by the complementary increase in the denudation rate in the surrounding heights. When we see, as in our own *Mittelgebirge* (in particular the *Erzgebirge*), that the surface systems of the individual steps, with their broad (*mulden*) valleys, are only weakly incised into the next system up, and that the incised *Kerb*-type valleys of the youngest part of the system cut back from the mountain foot right through all the step systems into the *dellen* region, then there appears to be a strong possibility that climatic–geomorphological causes have been operative, as Büdel (1935) has pointed out. Such a landscape is certainly not explicable by reference to crustal movements alone.

A long time ago Passarge suggested that fundamental research into present geomorphic processes was necessary in order to distinguish between present-day landforms and ancient features. Most geomorphologists would agree with this viewpoint. However, the individual processes cannot be considered in isolation as they act together in a most complex way.

One thing is of general validity in all this: material is moved by these processes within the limitations of the earth's gravity field and there would be no geomorphic change without such transportation. We can deduce geomorphic laws from this, for all exogenic processes produce work and so use up energy; they have, in fact, an 'energy budget'. An overall, somewhat qualitative, view of the energy budget has been attempted in the preceding statements on the varying balance between general surface denudation and linear erosion in various climates. The method employed here emphasises closer estimation of the gradient conditions of river channels. These gradient conditions give a true and relatively easily understood picture of the energy consumption of streams to which must be added the total output of surface denudation. Von Wissman has clarified matters to some extent in his work on the relationship between river load and lateral erosion, and it has been the aim of this paper to apply his findings to the basic questions posed by fluvial erosion landscapes. It is only on such a basis that insight can be gained into the complicated interplay of present-day erosion processes. The individual features of this picture conceal the special geomorphological problems but, basically, two major groups of causal factors require understanding: we are dealing either with the effects of changes in climatic–geomorphological conditions or with the features arising from earlier or present-day crustal movements or with a combination of both groups. Davis's theory of the erosion cycle and its later refinements implied that crustal movements alone are the most important factor in the formation of fluvial erosion. This fails to give due recognition to the fact that climatic changes occur which cause a change in landscape character, without crustal movement being in any way involved. This one-sided nature of the older ideas makes recognition of the vital importance of climatic–geomorphological factors all the more imperative.

REFERENCES

BORNHARDT, W. (1900). *Zur Oberflächengestaltung und Geologie Deutsch-Ost-afrikas*. Berlin

BÜDEL, J. (1935). Die Rumpftreppe des westlichen Erzegebirges. *Verh. und Wiss. Abhande. d. 25 Dt. Geographentages zu Bad Nauheim*. Breslau, 138–47

—— (1948). Die klimamorphologischen Zonen der Polarländer. *Erdkunde*, 2, 22–53

DAVIS, W. M. (1899). The geographical cycle. *Geographical Journal*, 14

HETTNER, A. (1921). *Die Oberflächenformen des Festlandes*. Leipzig and Berlin

JESSEN, O. (1936). *Reisen und Forschungen in Angola*. Berlin

KREBS, N. (1933). Morphologische Beobachtungen in Südindien. *Preussen Akademie d. Wissenschaft, Sitz. d. physikalisch-mathematischen*, Klasse 23

LAWSON, A. C. (1915). The epigene profile of the desert. *Univ. California Pub. Geol.*, 9, 23–48

PASSARGE, S. (1912). Physiologische Morphologie. *Mitt, Geogr. Ges. Hamburg*, 26

PENCK, A. (1919). Die Gipfelflur der Alpen. *Preussen Akademie d. Wissenschaft, Sitz d. physikalisch-mathematischen,* Klasse 17

PENCK, W. (1924). *Die morphologische Analyse.* Stuttgart

POSER, H. (1932). Einige Untersuchungen zur Morphologie Ostgrönlands. *Medd. om Grönland,* 94

—— (1936). Talstudien aus Westspitzbergen und Ostgrönland. *Zeitschr. für Gletscherkunde,* 24

SOLCH, J. (1935). Fluss- und Eiswerk in den Alpen zwischen Ötztal und St Gotthard. *Petermanns Mitt.,* 220

SPREITZER, H. (1932). Zum Problem der Piedmonttreppe. *Mitt. Geogr. Ges. Wien,* 75, 327–64

TROLL, C. (1947). Die Formen der Solifluktion und die periglaziale Bodenabtragung. *Erdkunde,* 1, 162–75

—— (1948). Der subnivale oder periglaziale Zyklus der Denudation. *Erdkunde,* 2, 1–21

WISSMAN, H. VON (1951). Über seitliche Erosion. *Colloquium Geographicum,* 1

10 The Theory of Savanna Planation

C. A. COTTON

THE SAVANNA 'CYCLE'

THE inselberg landscape (*Inselberglandschaft*) as described by Bornhardt (1900) has also been called the 'savanna landscape'; and the concept of a 'savanna cycle' has been introduced in an attempt to explain its origin (Cotton, 1942(a)). This landscape type, which is common in the tropics, is characterised by a great extent of very smooth, nearly level plains which are flooded in the wet season. From the plains inselbergs rise abruptly, forming very steep-sided isolated hillocks and hills, even mountains, of more or less bare rock.

Some examples of inselberg landscapes in a broader sense (Passarge, 1928), with less truly level plains though in other respects they superficially resemble the savanna landscape, are found in regions that are now semi-arid, where they have been either developed or at least much modified in form by the pedimentation process; these are better described as pediplains with residual inselbergs. In contrast with these, savanna landscapes, or inselberg landscapes in the strict sense, seem to have originated in hot climates that were more humid, notably in the climate that now prevails in the savanna belts ('humid savanna' of Büdel, 1957(a)), in which development of this landscape type seems to be still in progress. In these belts, equatorward of the desert, and semi-arid zones (including Büdel's 'desert savanna' and 'thorn savanna') the climate is hot all the year round and has alternating wet and dry seasons, the latter relatively short but sufficiently well marked (about three months in length) to interfere to some extent with the growth of forest. Büdel (1957(b) and (c)), who has discussed the origin of the savanna landscape, concludes that it develops also in the moist equatorial climate (with shorter dry seasons) that encourages the growth of rain-forest, and in the monsoon-forest regions as well. Significantly, Büdel (1957(a)) now maps the equatorial forest and humid savanna zones as one. He divides the Afro-European segment of the Northern hemisphere into climatic zones as follows: 0° to 12° N, tropical rain-forest and humid savanna; 12° to 16° N, thorn savanna; 16° to 18° N, desert savanna; 18° to 32° N, desert; 32° to 34° N, desert steppe; 34° to 36° N, steppe; 36° to 45° N, evergreen Mediterranean forest; 45° to

69° N, summer-green deciduous forest; 69° to 74° N, tundra; 74° to 77° N, frost debris zone; 77° to 90° N, glacier zone. In all such regions high temperature prevails at all seasons.

Tricart and Cailleux (1955) claim that savanna landscapes in general are *polygenetic*, that is, partly relict assemblages which have been developed for the most part by pediplanation in an earlier semi-arid episode or episodes which affected equatorial regions in the Quaternary. They admit that in the savanna belt plains have been thoroughly smoothed by the washing of fine silt from the higher parts of (presumably) pediplaned surfaces into hollows during the present-day phase of relatively humid climate; but they deny the activity particularly in hot forest regions, but also in savanna regions, of vigorous erosion processes such as could lead to planation. They claim indeed that, though chemical weathering is very active, erosion is now almost at a standstill in the equatorial belt except for the shaping of 'sugarloaves' of crystalline rock, existing relief, necessarily relict, being conserved practically without any modification. According to Büdel's theory, on the other hand, the processes of savanna planation are indigenous to hot-humid climates and have not been interrupted by Quaternary changes of climate. In other words, the savanna–plains landscape is *monogenetic* and a uniformitarian approach to its study seems therefore to be justifiable.

EXAMPLES OF ACTIVE SAVANNA PLANATION

Savanna planation seems to be active on the plain bordering the Gulf of Guinea at Accra and on the vast Wute Plain (altitude 400 to 600 metres) in the Central Cameroons (Clayton, 1956). The latter plain, as described by Sapper (1935), following. Thorbecke, and by Clayton, abuts at the rear against a retreating scarp, 400 to 600 metres high and 112 kilometres long, bounding the Ndomme Plateau to the north. Both these savanna plains are in the Gulf of Guinea 'refuge' in which, as suggested by Tricart and Cailleux (1955), the equatorial rain-forest survived through the Quaternary, thus excluding the alternative theory that planation took place during, and may have been restricted to relatively arid episodes.

THE MECHANICS OF SAVANNA PLANATION

Büdel (1957(b)) has noted that the rivers on a developing savanna plain are not normally incised in it even in the shallowest of valleys, but flow over its surface, or rather flood in the wet season, in ever-shifting courses. Allowance being made for an enormous difference in scale, the pattern of such rivers and their mode of operation recall the braided stream courses on a rock fan. The latter, as pointed out by Johnson (1932), are constantly shifting and eroding laterally, so that they wear the rock surface down evenly in the shape of a fan. Such, according to Büdel's theory, is the mechanism of a slow downwearing and shaping of the surface to river gradients, in this case gentle gradients, on a savanna plain—a process which must, of course, be interrupted many times by the temporary deposition of fine alluvium. The rivers engaged in this process rarely encounter solid rock, and do not erode it anywhere, as their load does not include any coarse waste. They erode only by washing away fine particles of the red surface soil of an intensely weathered regolith whose very considerable thickness is maintained by progressive rock decay below (Kubiëna, 1935; Clayton, 1956). This washing, which takes place when, in the wet season, the water-table is high and streams overflow widely, is more or less analogous to the sheetwash invoked by King as the chief agent of pediplanation. Büdel remarks that it is the wash and not any kind of mass movement that conveys fine waste into the main rivers and thus lowers the surface of the plains. Clayton (1956) agrees that the plains surface is progressively lowered by erosion, but suggests a different mechanism for this, namely a succession of sporadic 'rejuvenatory incisions' of small stream valleys alternating with what amounts to peneplanation of the interfluves between them.

Büdel insists on the presence of a thick zone of material weathered in place beneath the smooth savanna plain (Fig. 10.1). Such deep weathering is denied by Tricart (1958) in a review of Büdel, but it has been reported by Clayton, who has also quoted Thorbecke's observation of 'many metres' depth. In the opinion of Dresch (1953), the cover of fine red soil overlying bedrock under savanna plains in the western Sudan has not resulted from alteration of the bedrock. He claims that it is an alluvial deposit that has been brought by rivers from the south in Tertiary times and spread thinly over a very perfectly planed ancient surface, a pediplain from which inselbergs project up through the cover while the pediments around these are buried. Erosion, rather than further accumulation on the plain as a whole, is now in progress, but this is so slow as to be almost negligible. These plains are very nearly, though not quite, level, with a labyrinthine pattern of flood-time rivers in broad, shallow hollows, where the flow is very sluggish and liable to reversal. Though the Sudanese plains closely resemble those of the West African coastal region in some respects, their senile condition after survival for twenty million years or more (if this conclusion reached by Dresch is correct) renders them of little value, as compared with the other examples cited, for the study of processes that operate when savanna planation is active. Büdel describes, moreover, how the thickness of the regolith varies very much from place to place, especially where the lithology of the terrain is heterogeneous so that some of the rocks composing it yield more readily to rock decay and are therefore more deeply rotten than others. Thus beneath a regolith of great but very variable thickness (up to 30 metres and perhaps in places hundreds of metres) lies a very uneven basal surface of unaltered terrain (Fig. 10.1). It is

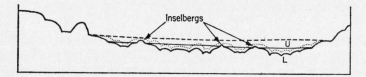

Fig. 10.1 Savanna planation at successive levels, with double surfaces of levelling: U, upper (subaerial) surface; L, lower (subterranean) surface or weathering front; in the latest cycle residual unweathered cores of rock have escaped planation and their outcrops have assumed the form of inselbergs, with great vertical exaggeration (after Büdel)

sometimes assumed (by the writer for example (1942)) that at least in temperate regions alteration of the nature of 'weathering' takes place only in a zone of aeration above the lowest dry-season position of the water-table; but, on the other hand, in the hot equatorial climate of Singapore some such alteration is, in the opinion of Alexander (1959), going on in a deeper zone that is continuously saturated with water and is thus never aerated—so that there is no downward limit. Probably because of the extreme rapidity with which she found chemical changes taking place, Dr. Alexander ignored the remote possibility that the zone of deep alteration she observed was in part relict from the late-glacial low stand of sea-level, alteration having thus begun in an aerated zone.

This concept of 'double surfaces of levelling', as Büdel names it, is important for the explanation of the numerous low inselbergs which are scattered through most savanna landscapes (Fig. 10.1). Some marginal inselbergs, including some of the highest, have originated as outliers of neighbouring or surrounding highlands which have become frayed at the edges as the plains have extended at their expense, making embayments in the highland margins (Fig. 10.2); this also takes place in the process of semi-arid pedimentation. Such high marginal inselbergs fringe the embayed eastern end of the Ndomme scarp, Central Cameroons. In contrast with these, however, ubiquitous, scattered, and generally smaller inselbergs are shaped by erosion from initial forms which have come into existence, according to Büdel's theory, when up-jutting unweathered salients of the unevenly decayed deep floor have emerged as a result

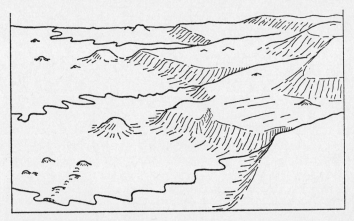

Fig. 10.2   Schematic representation of savanna planation in two cycles; the rear part of a savanna plain now developing (at the left) is separated by a scarp-foot pediment and scarp from a residual upland; rivers plunge over waterfalls at the front of the upland, which is not dissected but is bordered by the residual margin of the plain formed in an earlier cycle; high-marginal as well as lower-scattered inselbergs are shown

of a general ablation, or erosional lowering, of previously levelled savanna plains (see Figs. 10.1 and 10.2). Weathering and erosion have shaped these inselbergs ('shield inselbergs' of Büdel) into domical and tor-like forms. In some cases at least the erosional lowering of the plains surface has been intermittent, being caused, as Büdel postulates, by the entry of the savanna plains on new cycles of planation due to relative sinking of base-level, presumably caused in most cases by spasms of sensibly uniform upheaval.

Büdel's explanation differs but little from that proposed by Willis (1936), who also ascribes the durability of the cores that give inselbergs relative immunity from weathering of the rock types in them, and who envisages their survival as landforms through successive geomorphic cycles with an immensely long time span. This theory of polycyclic origin allows for growth to great height, provided an almost perfect immunity from erosion is conceded. According to Dresch, cited by Birot (1958), some Saharan monolithic domes have survived for tens of millions of years, increasing in relief through a succession of cycles. The polycyclic theory may afford an explanation of many of the high sugarloaf or domical inselbergs 'developed only in rocks of plutonic origin' to which King (1948) has attempted to restrict the name *bornhardt* (suggested by Willis in 1936 as a substitute for inselberg). The domical form of these is governed solely by their lithology and monolithic character; and Birot (1958) confirms the view (*see also* Cotton (1942(a)) and King (1957)) that such 'crystalline' domes are 'azonal', that is not peculiar to any climatic zone, being 'found from Arctic latitudes to the domain of the equatorial forest'. He thus throws doubt on the correctness of the claim made by Tricart and Cailleux that sugarloaf forms are produced by a special process of erosion peculiar to the equatorial rain-forest belt.

While over the whole extent of the savanna plains the land surface is, according to the theory of Büdel, being lowered by weathering *below* and by wash *at* the plains level, the area occupied by the savanna landscape is being enlarged by the marginal encroachment of pediments beyond the feather edge of the regolith of the plains on to the less thoroughly weathered, scarp-bounded rocks of inselbergs and also, more especially, of bordering uplands or highlands (Fig. 10.2).

It seems to be a mistaken theory that streams of water flowing along 'scarp-foot depressions' have hastened the sapping back of scarps either directly by lateral corrasion or indirectly by stimulating pedimentation along the scarp bases; the writer also favoured this theory (1942(b)). Linear scarp-foot depressions, though these are common, have been shown by Clayton to be in general minor features which, however they originate, are not and have not been river courses. Büdel ignores the possibility that lateral corrasion by rivers plays any part at all in cutting back upland scarps. Savanna plains are obviously enlarged, however, by retreat of the scarps that hem them in, and such scarp retreat must, if the theory of lateral stream planation be rejected, be due entirely to a process rather closely analogous to semi-arid pedimentation; but scarp-foot pediments, though present, are quite narrow and slope but gently—at a maximum of 3° according to Büdel and to measurements made by Clayton in Ghana. Though, as Clayton observes, the smoothness of profile may be broken by the presence of a problematical scarp-foot depression the narrow pediments merge generally into the savanna plains with decreasing slope.

It is the restriction of the pedimentation process to the margins of wash plains which may be of great extent that distinguishes this cycle of planation from the semi-arid cycle, in which pediments are much more widely developed, extending all the way down to the banks of streams into which the waste produced by erosion in the extension *and down-planing* of pediments is swept by sheetwash and by which this waste is carried away. It is, moreover, characteristic of semi-arid plains (pediplains) that 'even ancient and very flat (or nearly horizontal) pediments often show beneath a thin layer of transported material a smoothly cut, unweathered rock surface' (King, 1953). Semi-arid pedimentation, being, unlike savanna planation, a mechanical process taking place in a climate not conducive to the destruction of rock fragments by chemical weathering, is regarded as rapid in its operation (Tricart and Cailleux, 1955).

Marginal pediments that border a savanna plain, being composed of firm (if not entirely unweathered) rock with little or no regolith, may survive as residual benches if the plains surface is lowered in a new cycle following a sinking of base-level and a new scarp develops between upland and plains (Fig. 10.2). This, according to Büdel, is the explanation of piedmont benchlands.

THE TIME SCALE

The scarp descending from the Ndomme Plateau in the Central Cameroons is a feature on so grand a scale as to furnish an impressive reminder of the immensity of the time required for the savanna planation and accompanying scarp retreat that have produced the Wute Plain at its foot. Such a plain is enlarged only by retreat of the scarp at its rear, and the process of pedimentation at the foot of this Ndomme scarp appears to operate rather slowly; but however slowly the scarp as a whole retreats, its retreat keeps ahead of potential dissection both of the scarp and of the plateau above it such as would take place if consequent and insequent valleys were to develop vigorously. Dissection by such valleys would assuredly assume control if cool-humid conditions of climate like those prevailing in middle-latitudes were to supervene. The present scarp would then become an initial form for dissection in a 'normal' geomorphic cycle, just as slopes more or less similar to it but of tectonic origin which came into existence in the Quaternary in mobile regions (New Zealand for example) have been dissected (Cotton, 1958).

A high scarp like this must be immensely old in years, though when judged by criteria that might be applied in a cool-humid climate where dissection by streams is in progress it appears to be topographically young. Its youth is perennial, however, being due not to rapidity of development—though the question remains open whether the absolute rate of its development is or is not very slow—but to inhibition of the competing process of valley erosion. Such inhibition of valley erosion in the hottest parts of the tropical belt is well attested by various observers (Baulig, 1950; Tricart and Cailleux, 1955; Büdel, 1957(b)) who agree that its cause is the intensity at high temperatures of chemical weathering, which by breaking down rock fragments into fine debris deprives running streams of gravel and even of coarse sand with which to corrade their beds and so to deepen and grade their channels in solid rock (de Martonne, 1940). (Compare Fig. 10.2: the upland at the right is undissected, for, owing to inhibition of vertical corrasion, rivers plunge over the edge of the scarp as waterfalls.)

It is worth while emphasising the fact that inhibition of normal erosion is governed by a combination of high temperature and humidity, not by low latitude. It seems not to dominate all forested tropical regions to the extent implied by Tricart and Cailleux. In the Fiji Islands, for example, though in latitude 15° S, a forested landscape has been dissected by maturely graded rivers (Cotton, 1958). In the Samoan Islands dissection under forest is going on rapidly too (Kear and Wood, 1959).

Inhibition of the cutting of valleys by corrasion is a sufficient explanation of the dominance of scarp retreat in savanna planation, and it has not been suggested that the backwearing of scarps ever takes place rapidly. Rather is planation by this means far advanced, in regions that have continuously experienced hot-humid conditions and have also escaped recent upheaval, because of the vast length of the period during which it has been in operation. Glacial-age lowerings of temperature were never very strongly marked in the equatorial belt, amounting, according to Büdel (1957), to only 1° or 2° C, and there seems no doubt that in parts at least of this belt tropical weathering went on uninterruptedly and conditions favoured savanna planation throughout the Quaternary, as had probably been the case in pre-Quaternary times for millions of years. Büdel also remarks that it is of little consequence 'whether in an equatorial coast region the mean annual temperature is 26° or 24°'; the latter temperature will still be high enough to promote rapid rock decay under humid conditions.

According to Büdel the equatorial belt of tropical rain-forest increased in area in glacial ages, extending (with the closely associated savanna belt) to 15° N during the last glaciation of the northern hemisphere; but it may be necessary to limit the hypothesis of secular continuance of humid conditions from Tertiary through Quaternary to recent times to certain parts of the equatorial belt because Tricart and Cailleux cite evidence of a pronounced swing in glacial ages towards aridity in that belt. In interior Africa especially, according to their view, this may have caused interludes of semi-aridity or even in some places of extreme aridity, when forests must have been exterminated and semi-arid pedimentation—a rapidly acting process because of the mechanical erosion involved—must have supervened over large areas. Thus a history of secular savanna planation may be questioned in the case of some existing inselberg landscapes in low latitudes. The coast of the Gulf of Guinea has been suggested, however, by Tricart and Cailleux as one of a few 'refuges' in which the tropical rain-forest has survived through Quaternary arid episodes. If the suggestion is well founded this region, in which extensive savanna plains have been described that are still developing, must have escaped all episodes of arid or semi-arid erosion. It thus seems that these plains must be in a cycle of secular savanna planation.

ACCIDENTS AND INTERRUPTIONS

Besides plains of savanna planation that are at present developing there are other plains presumably of similar origin that are now youthfully dissected by numerous river valleys of moderate depth, perhaps as a result of upheaval. There are also many in various parts of Africa from which, according to Büdel's interpretation, the thick regolith normally underlying a savanna plain has been stripped away. Such removal may take place as a result of regional or local change to a cooler climate (which may accompany a lowering of base-level); and the same result may follow in the tropical belt without regional change of climate if a savanna plain is upheaved to become a very high and therefore cool plateau. Stripping away of the regolith by erosion at a rate so rapid that newly weathered debris is not produced as rapidly at the bottom of the layer may eventually lay bare the unweathered bedrock; the uneven surface of this will then be exposed so as to roughen the surface of the plateau.

A similar stripping must follow a local change of climate towards aridity, such as was brought about at the beginning of the Quaternary, when, according to Büdel (1954 and 1955), climatic zones similar to those of the present day were first established and the previously existing immensely broader zone of savanna and rain-forest climate contracted to a comparatively narrow belt near the Equator. Thus originated the desert plains of the southern Sahara and Somaliland (see also Kubiëna, 1935). Such an 'accident' introduces an entirely new morphogenetic system. Under an arid régime deflation by wind, with sorting of the now-desiccated material of the regolith into sand and dust, takes place rapidly. The sand being collected into ergs and the dust exported, it is possible for the underlying fresh rock to become exposed and to have its unevenness revealed, the salients of bedrock being now shaped into desert forms. Moreover, much dissection by valleys has also taken place in high-standing areas that are now arid—as in the Ahaggar region of the central Sahara. In those areas that are now the driest deserts, most of Quaternary time, with the exception of one or more pluvial episodes, has been available for these processes.

In pre-Quaternary times, on the other hand, there were presumably no deserts in Africa, the equatorial forest savanna belts being very wide. Extensive plateau remnants in the Ahaggar, for example, have been recognised as relics of late-Tertiary inselberg-dotted savanna plains, afterwards upheaved, on which the characteristic red regolith has locally been preserved under basalt sheets (King, 1951). The theory of the arid geomorphic cycle as developed by Davis (1912), which deduced the development of desert plains from initial landscapes of possibly strong tectonic relief, demanded an immense lapse of time during which average conditions favoured arid erosion —many millions of years (see also Cotton, 1942(b)). If there were no pre-Quaternary deserts in Africa, this rules out the concept of such desert planation. If, however, aridity has prevailed over large areas for some considerable fraction of a million years during the Quaternary it seems obvious that there has been ample time for the conversion of landscapes that had originated as savanna plains into typical arid-desert plains.

In North and South Africa initial surfaces on which the semi-arid pediplanation cycle as described by King (1951) afterwards developed may also have been savanna plains inherited from pre-Quaternary times.

TERTIARY PENEPLAINS

It is claimed by Büdel that savanna, or tropical planation not only accounts for existing inselberg landscapes, whether these remain as savanna plains that are still developing or have been upheaved or have become desiccated, but also has been the cause of the erosional development of so-called peneplains and piedmont benchlands in middle-latitude regions. This hypothesis was foreshadowed by Jessen (1938) and has been proposed also by Louis (1956). If it affords a correct explanation, these surfaces even if they were found undissected must be classed as relict forms which developed when (in the Tertiary) a tropical climate such as is now restricted to low latitudes extended far across the middle-latitude belts.

In middle-latitude regions the hot-humid climate of the Tertiary was eventually replaced by the Quaternary cool-to-cold climate. During glacial ages, and also in the less-cold interglacial and postglacial times, dissection by streams and in particular the cutting of deep valleys by rivers have radically reshaped the landscapes, destroying, according to Büdel, former widespread plains of savanna planation or reducing them to small remnants. These survive generally on major divides and on the interfluves between the branches of large rivers near their headwaters. Upheaval of the land has taken place to a greater or less extent in most places in the Quaternary, even where such uplift has not been great enough to produce mountain ranges, and the possibility of eustatic lowerings of ocean-level must also be considered as a stimulus to vertical corrasion. According to Büdel's theory, however, the almost ubiquitous deep incision of river valleys which has dissected the formerly planed surfaces, entirely destroying them in some regions, has been made

possible only by a change in the erosive capacity of the rivers themselves. Rivers in middle-latitudes flow now in valleys graded to gentle
gradients, so that farther and farther from the river mouths their
valleys are incised to greater and greater depths (until the ungraded
headwaters are approached). When, in earliest Quaternary times,
the climate became cool, the capacity to cut these deep valleys was
acquired by the rivers as a result of the supply to them of abundant
coarse waste such as they have since been making use of as tools of
corrasion, but which was formerly very scarce or absent, as it is
now in the equatorial belt of intense chemical weathering. The
availability for the first time of tools of corrasion thus led to the
dissection of middle-latitude regions, in particular Europe, bringing
abruptly to a close a widespread and long-enduring inhibition of
valley cutting. As regards time available, the million years of
Quaternary time seems adequate for the dissection of Europe, if the
tempo of erosion was anything like as fast as it is known to have been
in New Zealand.

As the hypothesis of peneplanation by downwearing of the land
surface presupposes a thorough dissection of the landscape by valley
cutting in a mature stage of the geomorphic cycle, acceptance of
a theory of inhibition of valley cutting and *therefore of dissection* in
pre-Quaternary times rules out the explanation of Tertiary surfaces
as true peneplains produced by downwearing. It compels the adop-
tion of some hypothesis that explains planation as due to progressive
destruction of residual uplands by backwearing. Just as in tropical
Africa the margins of plateaux are worn back so that they give place
gradually to new savanna plains, so in Europe in Tertiary times
upraised peneplains were gradually replaced by others graded to
lower base-levels, and this may have taken place without progressive
dissection. Alternative to savanna planation there is the suggestion
made by Tricart and Cailleux (1955) of semi-arid pediplanation.
Baulig (1957), who himself favours a theory of true peneplanation,
has pointed out, however, that late-Tertiary Europe, as indicated by
fossil floras, was warm and *humid* without signs of arid or semi-arid
episodes.

The validity of the foregoing explanation of the middle-latitude
landscapes as subject to savanna planation by backwearing at successively lower levels in the very late Tertiary, and as suffering
dissection by deeply cut valleys *only* in the Quaternary depends on the
correctness of Büdel's contention that not merely warm but really

ropical, hot-humid conditions persisted in Europe *until the end of the Pliocene*, when cooling took place suddenly, perhaps at the onset of the earliest Quaternary glaciation. A hot climate is generally believed to have prevailed in Central Europe in early-to-middle Tertiary periods, with the temperature 10° to 12° C above that of the present day (Woldstedt, 1954). Büdel (1957(c)), however, writes: 'Evaluation of upper Pliocene pollen analyses reveals that the forests then consisted of a mixture of Central European and evergreen trees—often to the astonishment of paleobotanists. The

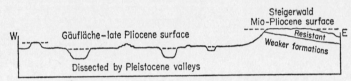

Fig. 10.3   Profile of surfaces of planation in Franconia, showing the Gäufläche and the Steigerwald (after profiles by Büdel)

conclusion may perhaps be justifiable that the Upper Pliocene hot climate of Europe resembled a subtropical monsoon climate'.

Peneplains in the Franconian basin are described by Büdel as typical extratropical examples of relics of plains of savanna planation that developed in late-Tertiary times. The most recently formed of these is preserved at an altitude of about 300 metres on broad interfluves between valleys of Pleistocene origin in the vicinity of Würzburg. Naturally the development of this level surface, the Gäufläche, in late-Pliocene cycles destroyed plains similarly developed in earlier Pliocene cycles at somewhat higher levels, with the exception of some marginal remnants; but owing to peculiarities of geological structure quite large areas are preserved of a much higher (and slightly warped) plain of considerably earlier date, namely Sarmato-Pontian (Mio-Pliocene). This forms, for example, a plateau on the Steigerwald at an altitude of about 480 metres, the boundary of which is a high structural escarpment (Fig 10.3). The preservation of parts of this higher plateau on resistant stratified rocks, where it has alternatively been described as a non-cyclic structural plateau, resembles that of the peneplain forming the Colorado Plateau and of the East Australian peneplain where it survives widely on the nearly horizontal Hawkesbury Sandstone of New South Wales.

Rising above the level of late-Pliocene planation in Franconia there are low monadnocks which are explained by Büdel (1957(c)) as inselbergs that projected up from a savanna plain. Above the level of Mio-Pliocene planation also there are monadnocks some of which are in this case truncated by relics of earlier-formed surfaces of planation, but the areal extent of these is not great.

In the Franconian basin relics of the late-Pliocene planed surfaces are at such uniform altitudes as to suggest that all the lowerings of base-level that have affected them have been eustatic. In other parts of Central Europe, however, there have been uneven upheavals, and flights of step-like benches of late-Tertiary age (piedmont benchlands) on the flanks of mountain massifs can be accounted for as a result of such updomings interrupting savanna planation.

REFERENCES

ALEXANDER, F. E. S. (1959). Observations on tropical weathering. *Quart. Journ. Geol. Soc.*, **115**, 123–44

BAULIG, H. (1950). *Essais de géomorphologie*. Paris, p. 63

—— (1957). Peneplains and pediplains. *Geol. Soc. America Bull.*, **67**, p. 925

BIROT, P. (1958). Les dômes cristallins. *Mémoires et Documents*, 6(a), (1), Paris CNRS, p. 33

BORNHARDT, W. (1900). *Zur Oberflächengestaltung und Geologie Deutsch-Ost-Afrikas*. Berlin

BÜDEL, J. (1954). Klimamorphologische Arbeiten in Aethiopien im Frühjahr 1953. *Erdkunde*, **8**, 139–55

—— (1955). Reliefgenerationen und plio-pleistozaner Klimawandel in Hoggargebirge. *Erdkunde*, **9**, 100–115

—— (1957(a)). The Ice Age in the tropics. *Universitas*, **1**, 183–91 (fig. 1)

—— (1957(b)). Die 'doppelten Einebnungsflächen' in den feuchten Tropen. *Zeitschrift für Geomorphologie*, **1**, 201–28

—— (1957(c)). Die Flächenbildung in den feuchten Tropen und die Rolle fossiler solcher Flächen in anderen Klimazonen. *Deutscher Geographentag Würzburg*, 89–121 (especially p. 100)

—— (1957(d)). Grundzüge der Klimamorphologischen Entwicklung Frankens. *Würzburg Geogr. Arbeiten*, Heft **4/5**, pp. 35 and 42

CLAYTON, R. W. (1956). Linear depressions (Bergfussniederungen) in savanna landscapes. *Geographical Studies*, **3**, 102–26 (especially figs. 1 and 11)

COTTON, C. A. (1942(a)). *Climatic accidents in landscape-making*, 2nd ed. Whitcombe and Tombs, Christchurch

—— (1942(b)). *Geomorphology*. Christchurch and London, p. 172

—— (1958). Fine-textured erosional relief in New Zealand. *Zeitschrift für Geomorphologie*, **2**, 187–210

DAVIS, W. M. (1912). *Die erklärende Beschriebung der Landformen*. Leipzig, figs. 142 and 143

DE MARTONNE, E. (1940). Problèmes morphologiques du Brésil atlantique. *Ann. de Géographie*, **99**, p. 112

DRESCH, J. (1953). Plaines soudainaises. *Rev. Géom. Dyn.*, **4**, 39–44

JESSEN, O. (1938). Tertiärklima und Mittelgebirgsmorphologie. *Zeitschrift für Erdkunde*, 36–49

Johnson, D. (1932). Rock planes of arid regions. *Geogr. Review*, 22, p. 660

Kear, D., and Wood, B. L. (1959). The geology and hydrology of Western Samoa. *N.Z. Geol. Surv. Bull.*, 63, 23–6

King, L. C. (1948). A theory of bornhardts. *Geog. Journal*, 112, 83–7

—— (1951). *South African Scenery*, 2nd ed. Edinburgh, p. 52

—— (1953). Canons of landscape evolution. *Geol. Soc. America Bull.*, 64, p. 743

—— (1957). The uniformitarian nature of hillslopes. *Trans. Edinburgh Geol. Soc.*, 17, 98–9

Kubiëna, W. L. (1935). Uber die Braunlehmrelikte des Atakor (Hoggargebirge Zentral-Sahara). *Erdkunde*, 9, 120–26

Louis, H. (1956). Rumpfflächenproblem, Erosionzyklus und Klimamorphologie. *Geomorphologische Studien (Machatschek Festschrift)*, 9–26

—— (1957 and 1959). The Davisian cycle of erosion and climatic geomorphology. *Proc. igu Regional Conference in Japan*, 1957 and 1959, 164–6

Passarge, S. (1928). *Panoramen afrikanischer Inselberglandschaften.* Berlin

Sapper, K. (1935). *Geomorphologie der feuchten Tropen.* Leipzig, 103–7

Tricart, J. (1958). Review of J. Büdel's Die 'doppelten Einebnungsflächen' in den feuchten Tropen, in *Rev. Géomorphologie Dynamique*, 9, p. 29

Tricart, J. and Cailleux, A. (1955). *Introduction à la géomorphologie climatique.* Paris, 155–217

Willis, B. (1936). *East African plateaux and rift valleys.* Washington (Carnegie Inst.) 119–27

Woldstedt, P. (1954). Die Klimakurve des Tertiärs und Quartärs in Mitteleuropa. *Eiszeitalter und Gegenwart*, Band 4/5, 5–9

11 An Alternate Approach to Morphogenetic Climates

WILLIAM F. TANNER

GEOMORPHOLOGISTS have varied widely in their willingness to use climatic indices. Some, like Lester King (1953), report more or less universal processes at work across many different climate types. Others profess to find genuine geomorphic differences between many climatic subdivisions.

MAIN CLIMATIC TYPES

The present writer occupies something of a middle position. He feels that, for the geomorphologist, there are four main climatic types, and that subtypes should be examined after these four have been firmly established. They can be listed as: wet; warm-dry (arid); cold-dry; and temperate, or moderate humid.

Important landforms associated with these climate types include: knife-edge topography; pediment and esplanade; glacial and tundra, and temperate. The fourth item in this list lies in between the other three, which occupy extreme positions (toward: high rainfall; high temperature; low temperature). This arrangement suggests a triangle, rather than a square, as the basic classification pattern. A square, for example, might emphasise the following combinations: hot-dry, hot-wet, cold-dry, cold-wet. However, the operational mechanics of the atmosphere are such that hot-wet and cold-wet are not readily distinguishable; there is, for all practical purposes, only 'wet,' with no great extremes of temperature, whether hot or cold. It is thought, then, that a three-cornered arrangement, with temperate in the centre, is superior to a four-cornered scheme (Troll, 1948).

PRECIPITATION AND POTENTIAL EVAPORATION

The parameters which have been chosen, for classification purposes, are precipitation and potential evaporation. Each of these is tentatively suggested as of greater significance, from a geomorphic point of view, than temperature. Furthermore, it is proposed that a suitable combination of these two parameters will also provide much, if not all, of the temperature information needed.

Both parameters have, however, wide variability. Potential evaporation may have any value from somewhere around 1 cm annually, to perhaps as much as 500 cm annually. And the annual figure for precipitation can be, apparently, anything from less than 2 cm to more than 2500 cm (certain stations in India). The relative value of 25 cm, where this is the order of the annual total, on the one hand, and where the annual total averages perhaps 1500 cm, on the other hand, is quite different. Hence, both parameters should be plotted on logarithmic scales.

Potential evaporation, as a climate indicator, possesses a disadvantage not shared to anything like the same degree by precipitation. It is, in effect still an uncertain factor, despite various methods of measurement (none of which has received unqualified acceptance). Nevertheless, since a general approach rather than a precise formulation is desired, the usual corrected values for potential evaporation are thought to be acceptable.

In order to test these ideas, the author collected precipitation and potential evaporation data from many parts of the world. Data for the United States before 1958 are easy to obtain (for summaries, see Linsley, Kohler and Paulhus, 1949). For the rest of the world, potential evaporation data are particularly hard to get. Those persons who very graciously collected information for this project included Shinjiro Mizutani, for Japan; Chester Wentworth and Jen-hu Chang, for Hawaii; DanYaalon, for Israel; and Arthur O. Fuller, for South Africa. Additional data were obtained from Penman (1954), for parts of Europe. It should be clear that those persons who were kind enough to collect data did so without necessarily endorsing the overall project.

The values so obtained were plotted on two charts; for most of the North American information, see Fig. 11.1, and for a broader presentation, see Fig. 11.2. In each case, potential evaporation (in cm per year) has been plotted against total precipitation (in cm of water per year). Fig. 11.1 shows how little of the total chart area is utilised by the selected American states. It is obvious that, for wide variations in the two chosen parameters, one will have to range farther afield. The other 36 states (excluding Alaska and Hawaii) fall within the area outlined by the 12 which have been plotted. It has been the purpose of this chart to cover essentially the full range of both P and E for the states shown, but on the other hand, it is recognised that a few exceptional figures may have been overlooked.

Furthermore, the method of approach has been to emphasise the variability within each state or area (hence a patch, outlined by a black line, rather than a point representing average values). The two dashed lines represent E = P, and E = 10 P, which approximate the values used by various workers as limits for the notion of semi-aridity (for a similar concept, see Senstius, 1958).

Various parts of the world are shown on Fig. 11.2. The circled area

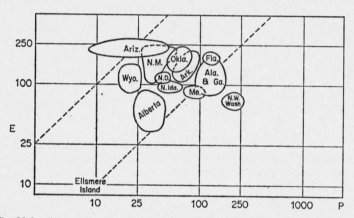

Fig. 11.1   Station data for 12 states, and scattered parts of Canada, plotted on a chart where the co-ordinates are precipitation (P) in cm per year, and potential evaporation (E) in cm per year. Each state, and part of the province of Alberta, is widely, but not necessarily completely, covered. The diagonal lines are E = P, to the right, and E = 10 P, to the left. They have been used, in some instances, as boundaries between humid, semi-arid and arid climates, but are presented here merely for information

for India covers those stations for which both P and E were available; the uncircled area, those stations for which E had to be estimated. Obviously, not all parts of India are included. The circled area for Hawaii is based on the best (numerical) information available. The tremendous variability of the parameters, within a fairly small land area, is plain. Representative states (Arizona, Wyoming, and Oklahoma) are included, without limiting circles, for comparison. The data for Arabia, north Canada, Korea, Borneo, Thailand, New Guinea, and Burma did not include figures for potential evaporation.

The basic information conveyed by Figs. 11.1 and 11.2 has been simplified and redrafted as Fig. 11.3. The external limits, taken from Fig. 11.2, are not thought to be final, and will undoubtedly require modification. The internal boundaries are shown as broad bands, rather than as sharp lines, and are labelled in terms of temperature. There are also probably quite a few exceptions to these temperature values. A few tentative subdivision listings are given, for purposes of illustration, but not as essential features of the chart.

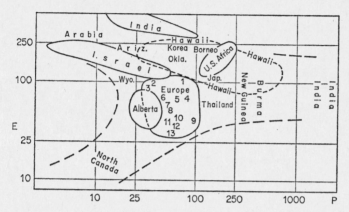

Fig. 11.2   Station data for four states (Wyoming, Arizona, Oklahoma and Hawaii), and various other parts of the world, plotted as in Fig. 11.1. Areas which are not circled indicate either (a) transfer of data from Fig. 11.1 or (b) incomplete information (in most cases, no data on potential evaporation). The numbers in the area representing Europe signify (1) Lisbon, (2) Athens, (3) Madrid, (4) Rome, (5) Belgrade, (6) Odessa, (7) Paris, (8) London, (9) Scotland, (10) Baltic Sea, (11) Moscow, (12) Leningrad, (13) southern Sweden

The present writer is not prepared to say that the four major types can be recognised in the field. Undoubtedly, in certain instances, recognition would be easy. In others, however, it might be much more difficult, or even impossible. This is due to the fact that climate, a complex entity in itself, is nevertheless only one of several variables in the overall picture. Any artificial scheme, where so many variables are involved, must have weaknesses. It is thought that the simplified chart given as Fig. 11.3 will have fewer weaknesses, morphologically, than more detailed climatologic diagrams.

Equilibrium landforms for some of these climatic types have been suggested by Cotton (1942). Our reluctance to recognise them, however, has stemmed from the fact that, in many parts of the world, the 'types' do not seem to be 'typical'. For example, the rounded knob topography (tors, *bornhardts*, etc.) which is supposed to identify the savanna also appears on many definitely non-savanna types (such as the selva, the pediment, and the tundra). This is because the equilibrium form is not the product of climate alone. The materials available must be considered also. Accordingly one of their most important characteristics may be described under the term directional weathering, for the basic facts underlying any climate-versus-landform scheme are mineralogical.

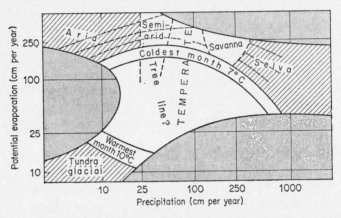

Fig. 11.3   A climatic generalisation of the station data shown in Figs. 11.1 and 11.2, with approximate boundaries indicated for: warm climate (coldest month 7°C), cold climate (warmest month 10°C), and the treeline. The class labels represent a geomorphic interpretation. It is not felt, however, that rigorous conclusions can be drawn from the data; boundaries shown are not precise, but are included for information because of the wide use of boundaries of this type in the past

ISOTROPISM

In the sense of the preceding paragraph, rocks are either isotropic or aelotropic. This result obtains obviously and directly. It can be observed, in the field, on a small scale. Many blocks, boulders, and smaller rock fragments show weathering–banding which indicates that they are isotropic (weathering has proceeded equally rapidly in all directions). This phenomenon has been observed by practically all field geologists, on both curved and non-curved surfaces. As a general rule (but not invariably) there is no visible bedding, schistosity or foliation present. Such weathering bands typically maintain a constant 'width' (that is, depth from the rock surface over a single block or boulder, and from boulder to boulder in the same lithology). Isotropic rocks produce *bornhardts*, tors, monoliths, sugarloaves among other features. They are not limited to the savanna. The present writer has examined features of this kind from many parts of the United States where savanna conditions do not prevail.

Not all rocks are isotropic to weathering, however. Many examples are known where structure and/or stratigraphy control the weathering. Such rocks are aelotropic. As they weather in preferred directions they produce mesas, buttes, *cuestas*, hogbacks, and other similar

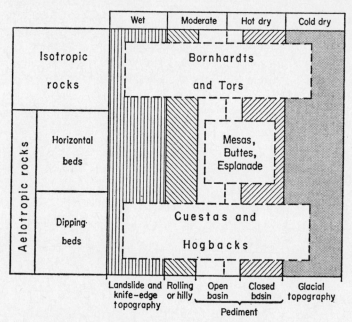

Fig. 11.4 Plot of rock isotropism against the four main climate types (at top of chart). Vertical shading and notes at bottom indicate isotropic effects

forms. Two main classes of aelotropic rocks appear: those where beds or planar masses are horizontal, and those where beds (or planar masses) are inclined. The former give rise to the esplanade, mesas, and buttes of dry regions, and the latter to *cuestas* and hogbacks. Inclined aelotropic rocks, under savanna conditions, develop topography in the form of *cuestas* much like they do elsewhere.

The general relationships between climate and directional variability are given, roughly, in Fig. 11.4. This chart is not comprehensive, and the division lines shown in it are not sharply or narrowly drawn.

Some distinctive landforms, such as dunes, are omitted because their development is not controlled by the common climatic parameters described in terms of moisture and temperature. Instead, they reflect availability of the right materials (sand-sized particles), plus windiness, and therefore constitute a special category. The various dune types have, however, equilibrium (depositional) forms. Fig. 11.4 likewise does not include coastal features. The latter are particularly complex, inasmuch as practically all forms of energy operate there. A detailed discussion of some of these problems has been presented elsewhere (Tanner, 1959, 1960).

REFERENCES

COTTON, C. A. (1942). *Climatic accidents in landscape-making*, 2nd ed., Whitcombe and Tombs. Christchurch

KING, L. C. (1953). Canons of landscape evolution. *Bull. Geol. Soc. of America*, **64**, 721–52

LINSLEY, R. K., Jr., KOHLER, M. A., and PAULHUS, J. L. H. (1949). *Applied Hydrology*. McGraw-Hill, New York, pp. 93–104, 156

PENMAN, H. L. (1954). Evaporation over parts of Europe: *Assoc. Internationale d'Hydraulique Scientifique*. Assemblée Générale de Rome, 3–4, 168–76

SENSTIUS, M. W. (1958). Climax forms of rock weathering. *American Scientist*, **46**, 355–67

TANNER, W. F. (1959). The importance of modes in cross-bedding data. *Jour. Sed. Petr.*, **29**, 221–26

—— (1960). Expanding shoals in areas of wave refraction. *Science* **132** (14 Oct.), 1012–13

TROLL, C. (1948). Der subnivale oder periglaziale Zyklus der Denudation. *Erdkunde*, **2**, 1–21

12 Area Sampling for Terrain Analysis

LOUIS C. PELTIER

AREA sampling is most useful in reducing the time and effort required for studies of large areas. It permits the formulation of rapid approximations and, because time is saved, it enables more detailed observations to be made and a greater number of variables to be considered.

Because this method of study is suited to observations in terms of measurements, the information so derived is in appropriate quantitative form for the comparison of widely separated areas. Thus, it aids in the elimination of biases of translation and recollection.

OPTIMUM TYPE OF SAMPLING

The optimum type of sampling is best determined by the problem and by the kind of answer sought. If a frequency distribution or probability expression of total characteristics is sought, a sampling of single points or small areas appears best. Effort is decreased, without a corresponding loss in accuracy, by decreasing the area of the observation points and increasing their number.

Detailed field observations on areas 50ft (5·2 m) square seem to be satisfactory for obtaining statistical data on topography, soils, vegetation, and parent rock material. Such quadrats are small enough to be easily handled, can be recorded in a single photograph, and usually present an essentially homogeneous milieu.

In a situation requiring reliance upon secondary information, such as an exploration of the relationships of mean slope, relief, and drainage texture with lithology, as the writer has had occasion to do for Cyprus and Fort Sill, Oklahoma, quadrats of one square mile seem most satisfactory. On Cyprus, an area of 3500 square miles, 60 to 70 samples were adequate and gave a satisfactory picture of the common situations. Secondary data recorded at scales smaller than 1:63 360 are generally poor and are not likely to reveal new relationships.

Fig. 12.1 shows a comparison of results obtained from point and transect samples. As is evident, the same general conclusions concerning frequency distribution are produced by either method.

It seems reasonable to suppose that there are observational errors of at least 1°. If direction, or the sequence of events involving movement, are important to the problem, then transect sampling appears best. This type of sample provides data for probability expressions of such things as the chances of finding an outcrop or drainageway, the chances of being able to see for 10 miles, or the chances that a motorised vehicle will get stuck. It is

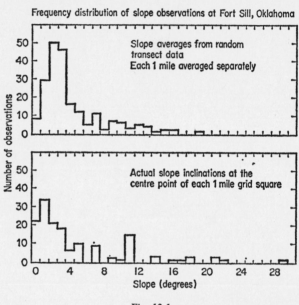

Fig. 12.1

particularly useful in exploring the probable association of two or more kinds of terrain situations. However, this type of sampling has proven most useful when combined with a form of gaming or simulated operational experiment.

A comparison of the results of random and systematic sampling of local relief at Fort Sill, Oklahoma, is shown in Table 12.1. The complete census provides a standard to be approximated. All samples, even the five per cent sample, define the mode and the general shape of the frequency distribution. The selection of the sample pattern depends upon the kind of answer sought. If a frequency distribution which describes an entire area is the objective, then a random distribution of sample points gives the best results

Table 12.1   Comparison of percentage results of sampling local relief at Fort Sill, Oklahoma

Local relief in metres	100 per cent sample (complete census)	50 per cent sample (regular pattern)	20 per cent sample (A) (random)	20 per cent sample (B) (random)	10 per cent sample (regular pattern)	5 per cent sample (random)
	%	%	%	%	%	%
230–242	0·6	0·0	0·0	0·0	0·0	0·0
218–230	0·0	0·0	0·0	0·0	0·0	0·0
200–215	0·0	0·0	0·0	0·0	0·0	0·0
185–197	0·0	0·0	0·0	0·0	0·0	0·0
168–182	2·6	5·2	3·3	3·3	0·0	12·5
153–165	2·0	1·3	3·3	6·6	6·6	0·0
138–150	0·0	0·0	0·0	0·0	0·0	0·0
123–135	2·6	2·6	3·3	6·6	0·0	0·0
108–120	4·7	3·9	3·3	3·3	13·2	0·0
93–105	6·7	3·9	0·0	9·9	0·0	0·0
78–90	8·7	10·4	6·6	6·6	6·6	0·0
63–75	7·3	7·8	9·9	3·3	6·6	0·0
48–60	2·6	2·6	3·3	3·3	6·6	12·5
33–45	12·0	13·0	9·9	16·5	19·8	12·5
18–30	45·5	41·6	46·2	36·3	46·2	50·0
0–15	4·7	5·2	9·9	3·3	0·0	12·5

with the least effort. However, if a map or other type of area distribution and regional stratification is the objective, then a regular pattern of samples appears to be best. Wherever the objective can be expressed as a question to be answered by the terms yes or no, sequential sampling seems most efficient.

OPTIMUM AMOUNT OF SAMPLING

Field geology has, in the past, been commonly based on the conclusion that data are where you find them. Krumbein (1959), in his discussion of non-orthogonal polynomial analysis, has shown a way of handling information so collected. However, it seems likely that there is an optimum point in the collection of such data beyond which an added increment of data does not yield a correspondingly large added increment of accuracy to the total conclusions. At least there seems to be a point beyond which the added information is unnecessary or irrelevant. Too much data can be so overwhelming as to be almost as much of a limiting factor in research as is too little data.

SIMULATED OPERATIONAL EXPERIMENTS

One of the most useful devices in applied quantitative geomorphology is the simulated operational experiment. In essence, this is playing games on maps or making believe in the field. It consists of assuming an initial point of departure which may be random or rational, therefrom assuming an action, and then of following out that action with the accumulation of measurements of the magnitude and frequency of significant or critical conditions that are encountered. By repeating this operation from other initial points, or by varying the assumed action, a body of data revealing the frequency distribution of the chances of success or failure may be obtained.

SEASONAL AND LATITUDINAL CORRECTIONS OF SAMPLING

Fig. 12.2 shows a sample distribution pattern for the study of selected world-wide variances. It is designed to solve a particular kind of problem which involves the world-wide distribution of physical conditions, particularly their mean value and frequency distribution. For example, the question may be raised as to what would be the range of values which would enclose 80 per cent or 90 per cent of the situations.

Where seasonal variance is involved, one may sample simultaneously in space and in time, using random geographical co-ordinates and random date–time groups. However, where the area involved includes a wide variance in latitude, the process of sampling by geographical co-ordinates results in a poleward concentration of the samples. This bias is most simply corrected by assigning a

Fig. 12.2

meaning to a sample in proportion to its latitude. This meaning is taken to be the proportion which the area of one degree square at the sample latitude would have to the area of one degree square at the equator. This is expressed by equation (1) in Table 12.3. Thus, fractional samples are dealt with. The correction factor, F, is multiplied by the value of the sample, and the sum of F is taken to equal the sum of N, the number of samples, for statistical purposes. From samples so selected and corrected, the number of samples on any continent is made proportional to the area of that continent, with a mean deviation of 1·7 per cent and a standard error of estimate of 2·1 per cent.

Some of the general conclusions derived from such a sampling procedure, and based on the sample sites represented in Fig. 12.2, are shown in Table 12.2. These distributions emphasise the dominance of gentle landforms, the approximate balance between forested and unforested terrain, and the geographical importance of desert

Table 12.2 World-wide frequency distribution of physical conditions on land

Landform	%	Vegetation	%	Climate	%
Mountains	24·7	Ice-cap or barren	11·8	Tundra and frost	16·4
(more than		Tundra	7·9	Microthermal	19·5
320 ft/sq mile		Desert	7·5	Mesothermal	17·5
Hills	18·6	Semidesert	10·2	Desert	26·5
160/319 ft/		Steppe and prairie	11·9	Tropical	20·0
sq. mile		Woodland and		Total	99·9
Plains	46·7	cultivated	31·4		
0/159 ft/		Forest	18·4		
sq. mile		Total	99·1		
(Antarctica)	9·9				
Total	99·9				

and tropical climates as defined by Köppen. Indeed, the pleasant mesothermal climates appear to prevail over only a small part of the world.

INFERRING VALUES OF TERRAIN PARAMETERS

In many instances, map data in adequate detail are lacking for optimum sampling purposes. Even in the United States, the average scale of the best available topographic map is only 1:273 000; and in Europe, including the USSR, the situation is no better. The best information for much of the world is on a scale of 1:1 000 000. Therefore, for many places it is necessary to infer or estimate the topographic situation on the basis of what is known to be present elsewhere under comparable circumstances.

Such estimates seem best done by using local relief, the maximum difference in elevation per square kilometre, as an indicator. For crude data on local relief, a general relief factor—the maximum difference in elevation within a 100 square mile area—may be used. This is the area within a square measuring 10 mile on a side, or a circle with a radius of 5·6 mile. The general relief factor in such an area is related to local relief in equation (2), Table 12.3.

By the use of such an estimate of local relief, one may then proceed to guess at other values of terrain parameters. Average slope seems best approximated by using equation (3) in Table 12.3; and average height of valley walls seems best approximated by using equation (4) in Table 12.3.

Table 12.3 Equations representing significant relationships in area sampling

Equation (1) $F = \dfrac{69 \cdot 0 \; [69 \cdot 2 \; (\cos \text{latitude})]}{69 \cdot 0 \; [69 \cdot 2]}$

  where F = correction factor for latitude

Equation (2) $\log \overline{R_L} = 0 \cdot 08335 + 0 \cdot 79606 \log R_G$

  where $\overline{R_L}$ = mean local relief in ft/sq. mile

  R_G = general relief in ft/100 sq. mile

Equation (3) $\overline{S} = 0 \cdot 024 \, \overline{R_L} - 0 \cdot 66$

  where S = mean slope in degrees

  R_L = as above

Equation (4) $\overline{h} = 0 \cdot 286 \, \overline{R_L} + 11 \cdot 56$

  where $\overline{h}$ = mean height of valley walls in feet

  R_L = as above

Texture, or the spacing of drainageways, however, should, for general estimates, be dealt with only after the controlling type of climate has been determined; then, the best equation relating the mean distance between drainageways in a random transect to mean slope can be selected. These equations, which are graphed in Fig. 12.3, are:

$\log \overline{S} = 4 \cdot 23 \log \overline{G} - 1 \cdot 22$ (for a moderate or mesothermal climate)

$\log \overline{S} = 6 \cdot 0 \log \overline{G} - 2 \cdot 86$ (for a semidesert climate)

$\log \overline{S} = 2 \cdot 89 \log \overline{G} + 0 \cdot 04$ (for a desert climate)

$\log \overline{S} = 2 \cdot 04 \log \overline{G} - 0 \cdot 57$ (for a tropics climate)

where $\overline{S}$ is the average slope, and $\overline{G}$ is the average number of drainageways encountered per mile of transect. These estimates are, at best, crude, for within these general estimates there is a variance which appears to be related to the weathering characteristics of the rock, the permeability of the soil, and the effectiveness of vegetative cover.

The displacement, in Fig. 12.3, of the curve for a moderate climate from that for a desert climate may be interpreted as the result of increased rainfall and runoff. The further displacement of the curve for a semidesert climate, however, seems best interpreted as the result of a lesser rainfall and runoff than in regions of moderate climate, but with poor protection by vegetation. The divergent curve for tropical conditions suggests a different geomorphic process;

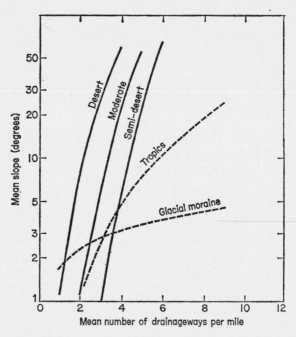

Fig. 12.3 The relationship of topographic slope and topographic texture for different climates

probably it is a reflection of landslide action on the steeper slopes. The curve for morainal topography also is divergent because it reflects the initial irregularity of the surface. In compiling statistics, kettle holes are included with drainageways in an attempt to measure the irregularity of the surface.

Many more such measures of the values of terrain parameters are needed, and many more systematic observations are essential. At the moment, the greatest need seems to be for the establishment of numerical scales to measure lithologic and soil characteristics as well as the effectiveness of vegetative cover, and for the correlation of these parameters with geomorphic measures.

REFERENCE

KRUMBEIN, W. C. (1959). Trend surface analysis of contour-type maps with irregular control-point spacing. *Journal of Geophysical Research*, **64**, 7, 823–34

13 Climatogenetic Geomorphology

JULIUS BÜDEL

THE aim of geomorphology is to order, and interpret the significance of, the immense variety of the earth's surface relief. This is only possible through a *genetic* examination. Further, since we have discovered that the deciding factors in the formation of relief are the exogenic processes steered by climate together with certain other factors, we propose the term *climatogenetic geomorphology*.

The earth's climate is divided into zones on a latitudinal (zonal) basis, on the basis of longitudinal position within the continents, and on altitude (as in Lautensachs, 1953). All these climatic zones are divisions of a continuum (Bobek, 1953) which contains few sharp breaks. Such do occur (for example the north and south limits of the Dinaric karst or of the Crimea) but, generally, a gradual change is typical. The change in relief-forms occurs parallel to that of climate. Both are similar changes, and the former is no less significant that the latter. An episodic river in the desert (wadi) has different characteristics from a perennial river in the savanna. Every grade of transition in river régime is found between these two. The same is true of morphological effects and of relief features produced by rivers all over the world.

This change in the exogenic forces due to climate predominates only on the broader scales (river lengths averaging 100 kilometres) On more local scales (distances as little as 10 kilometres) it is far outweighed by the more rapid spatial change of different, chiefly aclimatic influences. The two most important aclimatic factors derive from characteristics of the lithosphere on which the exogenic processes act. Additional ones arise from the combined influence of geological (endogenic) factors, climate and man himself. All these combined *aclimatic* factors are distinguished from those purely climatic factors defined below.

PETROVARIANCE

According to their hardness, surface rocks resist erosion by exogenic processes controlled by climate. The differences in relief caused by this are best perceived at medium scales. Along the same river, a valley wall of continuously similar height can, within a few hundred or even tens of metres be steeper or flatter depending on whether the rock is hard or soft. Often resistant layers can be seen standing out between soft ones like sharp ribs or prominent steps. Thus in many cases the effect of *petrovariance* can be grasped literally with the hand. In fact, some connection with petrovariance may be ascertained in the case of the majority of relief-elements of the earth. It is therefore not surprising that for eighty years—from the middle of the nineteenth until the beginning of the twentieth century—geomorphology was principally concerned with this question, and that even today many geologists see the most probable explanation of all relief-elements in the 'relativity of landforms to the rock', so that all other explanations are treated as unimportant.

After nearly a century of 'classical' morphology there is little need to mention that the influence of petrovariance on relief may be found *in some form* in almost all areas affected by erosion.

Accordingly, given that *all* forms of erosion (not forms of deposit!) are developed on a rock base of varying composition, the question is: How can it be assessed both qualitatively and quantitively? Both vary considerably and *the variation follows a pattern, depending on the climatic zones*. The ability of the rocks to withstand erosion does not depend on their physical but rather on their *morphological hardness*. This varies according to climate. Granite, for example, is one of the hardest rocks in the mid-latitudes whereas in the moist tropics, where deep weathering prevails, it is soft and sand-like.

In the polar regions (where perennially frozen ground limits karst development, see Corbel, 1957) chalk is by no means hard. In the mid-latitudes, (where there is a strong tendency to karst) chalk, due to the development of underground streams, is one of the more upstanding rocks. Therefore, there is a very clear difference in the morphologies of extratropical and tropical karsts despite identical rock composition (H. Lehmann, 1955).

EPEIROVARIANCE

As the second geological factor, epeirovariance determines the speed of vertical uplift or depression of parts of the crust, both at present and in the more recent geological past. (Traces of older more or less completed tectonic movements can be seen extensively on the earth's surface in petrovariance; they are therefore also taken into consideration in our system.) Epeirovariance—a term which has been stimulated by the ideas of W. Penck (1924)—is the main factor determining whether an area is regarded as one of subsidence governed by forms of deposition or as one which is uplifted and governed by forms of erosion. Epeirovariance, then, governs the ultimate height and general steepness of slopes. Of course, the form of an uplifted block or a step produced epeirogenetically appears in its pure form only in the relief of the ocean bed. On land, the effect of exogenic erosion processes begins at once, changing the 'raw block' (for example the Harz mountains) into a greatly reduced, many-membered relief of flat surfaces, ridges, slopes and valleys which we see today. From this we draw the general rule: the greater the speed with which the crust is uplifted, remaining so for a considerable period of time, the greater is its flattening and change in shape due to exogenic factors. The reason for this lies above all in the rate of erosion, which increases with each steepened slope; it lies also in the fact that the higher a mountain is raised so is it lifted into a climate of more extreme weathering and erosion, until its peaks rise above the treeline and eventually above the snowline. It is, however, immaterial what percentage of an uplifted area is *in statu nascendi* at any time or whether this is the result of subsequent erosional planation; for epeirovariance determines the lateral extent of the uplifted area, that is the shape of mountains in plan view: the rhomboid shape of the Harz, the triangle of the Massif Central, the arc-shape of the Alps.

DISTANCE FROM BASE-LEVEL

Let us imagine the theoretical case of a raised block such as the Harz mountains, the Central Plateau or something of similar form the size of the island of Borneo, and that all parts of it, at one particular time, were raised equally and that all other attributes (petrovariance, climate) did not change over the whole tabular area of the block. This block, both in the centre and on the periphery, takes on a new relief form according to the distance of each part from the surrounding base-level of erosion. Near this level, we should find short, deep and narrow valleys of the small coastal rivers, leading down steeply to the sea; and next to these broad valleys with comparatively gentle banks, about the lower courses of the larger rivers, whose upper courses reach well into the hinterland of the uplifted block. There, near the water divides, only smaller rivers are found with, of course, less total over-deepening (although, in the case of rapid headward erosion their sides may be just as steep as those of the coastal rivers). It should be noted that in the idea embodied in the statement 'the parts of the lower course of such river systems which have been caused by headward erosion must be *older* than the branches of its upper course or the subsequently formed short coastal rivers', one is reminded in part of the ideas of W. M. Davis about 'young' and 'mature' valley forms.

THE INFLUENCE OF HIGH RELIEF

This sums up another group of factors which is also connected with epeirovariance, but is more clearly determined by climate. It can best be considered as divided according to the effects of denudation and of erosion. The latter effects are influenced by the secondary effects of climates in distant parts of a catchment (external climate influence).

Denudation effect

Let us take an incline of medium slope—for example about 22°. At each point on it there is a certain proportion of accumulation and erosion, and in the case of the latter there is also a balance between the proportion of material brought from the higher slopes and that transported away down the slope. But even in cases of uniform gradient this balance may be altered if the incline becomes steeper at the top (concave bend in the incline), or becomes flatter or even level at the top (convex bend). In the first case an increase in the material brought from above can slow down the rate of reduction of the flatter incline below. In the case of the convex bend, an increase in steepness and erosion of the lower part of the incline, perhaps by incision, may lead to a complete 'removal by denudation', that is to a reduction in the rate of removal of the upper relative to the lower slope. Observations about this were made by Büdel (1961a) and Wirthmann (1962) on the Stauferland Expedition.

Effects of erosion

In the same way the higher parts of a valley can influence the lower parts. Bodensee collects the debris brought down by the Rhine and so the lower part of the river takes on a completely different character from the upper reaches. Glaciation in mountainous regions can cause heavily charged rivers to enter lower regions; according to the width of the valley and gradient of the *talweg* these may produce completely different effects from an autochthonous channel. This factor may be almost as important as those of zonal, large-scale effects.

Climatic influence of distant regions

Distant mountains receiving a lot of rain can cause permanent rivers (Nile, Niger) to flow through deserts and produce direct and indirect morphological effects of a kind completely different from those of the local desert wadis. The restriction of dune development is one example of the indirect effects of such permanent rivers.

HUMAN INFLUENCE

This term encompasses all influences on the existing balance of the denudation processes arising from the action of man. In many parts of the earth at different periods of time these effects have been just as important as those due to large variations of climate.

CLIMATIC VARIATION

The five factors already mentioned would still have their effect on relief formation if there were a uniform climate on earth. (Only the influence of high relief requires a high altitude climatic type—including the far-reaching influence of moist mountains.) The influences arising from the *differences in the climatic zones*, which we call 'climatic variance', are superior to these five aclimatic factors, however.

Under certain climatic conditions the aforementioned factors may change their valence; rocks may change their position in the scale of 'morphological hardness', mountains which are equally sharply uplifted may be depressed in different ways, final stages in down-wearing toward base-level may be reached earlier or later, and the influence of high relief and particularly human influence may take on a completely different form. The reason for this lies in this second, decisive fact; at each point on the earth, climate determines the prevailing combination of exogenic or morphogenetic processes. It is these which *actively* form the relief: the previously noted factors are only rather passively contrary influences (petrovariance and epeirovariance) or subsidiary influences (distance from base-level, high-relief influence and human influence). Thirdly, factors originating in tectonic movements whether older, long-past or young ones still in process (petrovariance, epeirovariance, distance from base-level and high-relief influence) and above all, influences due to man, are distributed over the earth irregularly and at random.

So the six grades of morphological hardness, under which the

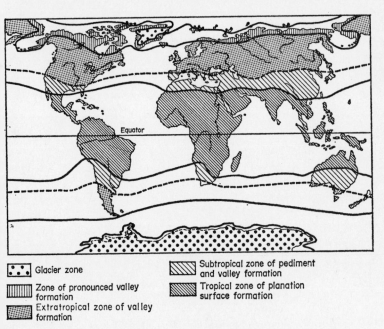

Glacier zone

Zone of pronounced valley formation

Extratropical zone of valley formation

Subtropical zone of pediment and valley formation

Tropical zone of planation surface formation

Fig. 13.1   The present-day climatomorphological zones of the Earth

main rocks of the earth's surface may all be included, are repeated indiscriminately from Pole to Equator in a varying pattern: chalk, dolomite and marl; sandstone; *flysch* series; plutonic and volcanic rocks; crystalline slate or schist; and young Tertiary and Quaternary sediments. The composition of Germany's mountains (and also their proportion of flat surface) is similar to those of North Africa, Asia Minor or eastern Australia. Opposed to this, the earth's climatic zones show a regular pattern corresponding to that of the major plant formations (and even, to a considerable degree, the overall pattern of land use). These direct and indirect effects of

climate together determine the particular combination of morphogenetic processes (weathering, soil formation, erosion and deposition)
in any specific part of the earth. In this way, morphogenesis forms
part of the great continuum of regular zonal change across the
globe. The prominence or even dominance of the climatic factor
and its uniquely consistent relationship to morphogenesis makes
geomorphology part of the science of geography.

CLIMATOMORPHOLOGICAL ZONES

According to the effects in each climatic region of this complex of
active morphogenetic processes, the earth's surface may be divided
into the following climatomorphological zones (Fig. 13.1, Büdel,
1948, 1961b).

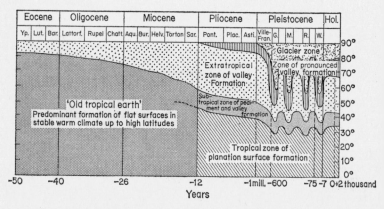

*Fig. 13.2 Climatomorphological zones of the northern hemisphere in a meridional
belt from the Equator to the North Pole, through Central Europe—since the early
Tertiary*

(1) Glaciated zone (polar regions and high mountains).

(2) Zone of pronounced valley formation (in polar regions
which are today unglaciated; notably underlain by permanently
frozen ground).

(3) Extratropical zone of valley formation (most of the
mid-latitude regions). Valley-forming processes are of little importance today but fossil features of glacial periods are usually much
in evidence (Fig. 13.2).

(4) Subtropical zone of pediment and valley formation (zone of transition between (3) and (5); for naming of this zone see Mensching, 1958).

(5) Tropical zone of planation surface formation.

The customary, classical division, which goes back to A. Penck (1910) and W. M. Davis (1912), of climatic influences on relief only recognised glacial, humid and arid zones (with forms in between these, such as semi-humid, semi-arid). In reality, however, this system was not derived from the total complex typical of the main climatic zones of current morphogenetic processes but merely from one part; basically it names only the main forms assumed by surface water. Therefore, our system is in agreement with the 'classical' one only in the morphological zone termed the subglacial or zone of glaciers. Within the remaining subaerial parts of the continents the division between humid and arid appeared to us less radical than the subdivision of the humid area between the Equator and polar inland ice into the above zones, according to the variation in the complex of formative processes. The steppes and deserts are arranged (in spite of their wealth of small distinctive landforms, see, for example, Mortensen, 1958; Wilhelmy, 1958; Knetsch, 1959), on the basis of their major forms—our types (5), (4) and to a certain extent also (3). Separation into a size order was not possible. The reasons for this lie in the climatogenetic view (see below) and particularly in the fact that some of the largest present-day deserts, such as the inner Australian (Bremer, 1962) and the Sahara (Büdel, 1963 and earlier), still bear widespread remains of fossil landforms due to former wetter conditions.

Of the subaerial zones of our divisions (2), (3), (4), (5), the second (the zone of pronounced valley formation) and the fifth (the tropical plain formation zone) represent the critical extreme cases. Strangely enough, however, the two have one common feature; weathering reaches in both cases well below the ground surface and surface erosion processes persist in previously etched linear courses. In both zones *deep weathering* takes place. This is an important preparatory process. As kinds of deep weathering vary considerably, so the features of erosion vary accordingly. This is why these two zones differ from the others. Zone (2), the zone of pronounced valley formation, lies in the high polar regions between the polar treeline and the snowline: therefore, it is associated with the tundra and frost rubble zones and particularly with perennially frozen ground. Today, this zone covers eight million square kilometres in the northern hemisphere and is equivalent to four-fifths the area of Europe (Fig. 13.1). During the Pleistocene period the area of tundra and frost rubble climate in the central belt of Eurasia, North America, Patagonia and New Zealand was almost double this extent (about 15 million square kilometres). It is difficult to draw the former equatorial limit of this treeless polar climate belt exactly, since it often merged with a loess–steppe and not a belt of forest; its extent is known fairly exactly only in western and Central Europe. While such climates prevailed in the Würm cold period about 50 000 years ago (when many features initiated in the Riss and earlier cold periods were reactivated), and in some short postglacial periods, the morphology changed little in some areas (for example in Central Europe).

Valley formation

In contrast, in formerly ice-covered polar and mountainous regions, the processes of excessive valley formation have only been at work in the short space of postglacial time so that fossilised traces of these past processes can also be seen clearly in Central Europe. The most important of these traces is *valley formation*. The valleys are usually steep-sided with terraced flanks and broad, slightly convex floors of rubble but with mostly a very even *talweg*. Even when bands of very hard rock are crossed there is only a slight increase in gradient (with a gentle narrowing of the valley floor) which never produces waterfalls or important rapids. Both are connected in the middle latitudes almost exclusively with the subglacial relief of formerly glaciated zones (Rhine Falls, Niagara Falls, the waterfalls of the Alps and northern Europe). Of the so-called periglacial area of the former tundra and frost-rubble zones, valleys with regular gradients and debris-covered floors which are excessively wide in proportion to the width of the present river are characteristic.

These profiles point to the fact that valleys, despite their broad floors, formed *quickly*. The generally steep valley sides attest the same. The terraces in these (for example in the Middle Rhine, the Moselle, Middle Elbe with the Saale and Moldau (Vltava), the Danube Gorge and many other stretches of the central German rivers with their frequently datable debris) also show that 80 to 90 per cent of the valleys cut into the high altitude surfaces of the German Mittelgebirge were deepened by stages in relatively short periods of time. While the Pleistocene cold phases together include not even half (about 350 000 to 400 000 years) of the total duration of the Pleistocene, valleys were deepened in this time by between 195 and 235 metres (for example the Moselle). This means, therefore, that during the glacial phases there was an average deepening of about 0·6 metres in 1000 years *over the whole width of the valley floor!* On the Main (and similarly on the lower Neckar) the river accomplished this a second time in the Pleistocene, thanks to tectonic uplift with respect to base-level.

Such results are further attested by the fact that the non-glaciated (periglacial) Pleistocene valleys of the German Central Uplands were able to reach back right into the core of the hills of the Black Forest, Rhenish Schiefergebirge or Erzegebirge (Büdel, 1935). The higher summit surfaces were removed or dismembered by the extension of lateral valleys, small parts only remaining intact. The highest reaches of the valleys—now dry—show the characteristic form of *dellen* which, with all their features (notably their asymmetry) can be recognised as diagnostic examples of the valley formation of the glacial phases. According to the results of the 1959–60 Stauferland Expedition to south-east Spitzbergen, a similar type of valley formation found there is not yet universally so advanced. The non-glaciated areas in south-east Spitzbergen were considerably larger than today throughout the postglacial climatic optimum (7000–500 B.C.). The present-day glaciation, which covers about 55 per cent of south-east Spitzbergen is essentially the combined product of a new depression of the snowline and land uplift in the last 3000 years. There are three reasons for this. The first is that these oceanic parts of the present-day Arctic were covered during the Würm glaciation by an area of inland ice which disappeared about 13 000 B.C. (Büdel, 1961a). A frost-rubble climate has subsequently predominated in these areas which have been ice-free for only about 15 000 years. The second reason is that the Holocene deepening of the valleys ran parallel to a particularly vigorous isostatic land uplift of 250 metres in the same period of time. Thirdly, this caused the lower parts of these islands not to emerge from the sea until late, which has exposed them to the present frost climate for an even shorter time, that is the lowest edges of the foreland, formerly 15 metres and less below sea-level (according to radiocarbon dating of whale bones embedded in the frozen ground by Dr Münnich, Physical Institute, Heidelberg).

The mode of valley formation and its results can be seen all the more clearly here in Spitzbergen since the initial form (subglacial relief) of the Würm is easily reconstructed; the isostatic response, the duration of frost climates in the Holocene and the currently active processes of erosion can all be measured and observed exactly.

Büdel (1961a) and Wirthmann (1962) examined about 25 valleys over 20 kilometres in length, whose tributaries for the most part do not rise from glaciers. Of these, *all the larger ones* have in recent postglacial times taken on a completely even transverse profile with a broad debris-covered floor. In ten of them a lowering 60 metres or 6·6 metres in 1000 years (or ten times as much as in the case of the rivers of Central Europe in the last Cold Period) has been produced since the higher sea-level of the 'Riegel series', 9000 years ago. This can be explained above all by the steepening of the regional slope of these comparatively short coastal rivers caused by recent isostatic uplift. The incision is due only to a small extent to the presence of soft Holocene marine sediments; for the most part it has developed in the underlying Mesozoic strata of marl, slate with bands of chalk and clay ironstone plus some hard sand-stones and arkose in between (somewhat similar to our Keuper marl).

Only a few of the valleys are different. The Stauferbach (12 kilometres long with its mouth at the steep south-west edge of Barents Island) makes its final descent in a series of steps arising from the outcrops of two large and compact basalt layers. The stream has not yet cut right through even the higher of these, and it tumbles over it in a waterfall. The resistant material, therefore, preserves in this valley the effect of the very rapid recent uplift. It demonstrates the 'normal' conditions of this zone. Above this waterfall (92 metres above sea-level), the middle and upper reaches of the stream are set in a trough-like valley which, unlike other valleys, has *gentle* banks; these rise from the rubble-covered floor in a uniform concave–convex profile to the remnants of old surfaces at a height of 400 metres (Friedrichsberg in the west, Hohenstaufen in the east, see Fig. 13.3). This situation makes it possible to follow the steady succession of erosion processes on pure solifluction slopes—without interruption by recent phases of incision. On all steep slopes (between 45° and 60°) there is rapid erosion in all climates, of course. However, in an area of frost climate there was found to be considerable erosion even in valleys with flat floors and gentle banks; field observations suggest about 15 metres in the last 15 000 years, or about one metre per century over the whole width of the floor and the smooth gently concave marginal banks. In other words, the Holocene deepening of the Stauferbach corresponds in manner and magnitude to the mid-course valley deepening in Central Europe during the glacial Pleistocene. It could well have been the most severe ever known and may even have exceeded the prevailing linear erosion of *riverbeds alone* in the

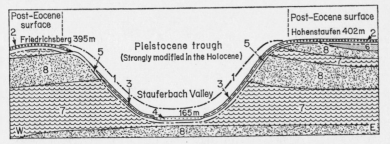

Fig. 13.3 The Stauferbach valley in Barents Island (south-east Spitzbergen): an example of pronounced valley formation in the present-day frost debris zone.

First relief-generation: post-Eocene surface presently at altitude of 400 m; second relief-generation: Pleistocene trough; 1—approximate valley profile at end of the Würm cold stage; third relief-generation: rapid valley development and slope backweathering in the Holocene; 2—cryoturbation mantle of the old surface; 3—solifluction mantle with drainage rinsing on the slopes; 4—present-day valley gravel; 5—realm of ground ice action, the upper part of the perennially frozen soils with severe frost disturbance and shattering of rock; 6—resistant arkose; 7—shales with chalk and clay-ironstone bands; 8—basalt and dolerite sheets

present deciduous woodland climate of Central Europe by several hundred, perhaps even several thousand per cent.

Permafrost factor in erosion

The mechanism of this erosion is dependent on the presence of permafrost, which is here about 350 metres thick. The role of the seasonal thaw (on an average 30 to 40 cm deep with extremes 8 and 67 cm) is important, for in the three months of summer many denudation processes are active, notably channelling of slopes poor in plant cover and processes associated with melting snow. The first of the new season's snow is laid down at the end of August and is steadily increased until May, with only few and brief periods of thawing. The annual thaw begins at the end of May and persists strongly throughout June. Many perennial snow patches and snow cornices still remain, however, in suitably sheltered places in July and August, or until the new season's snow begins to accumulate. The water from thawing ground-ice gathers on all the slopes and joins the snowmelt water. Not a drop percolates because of the frozen ground below. Evaporation is minimal due to low summer temperatures (absolute maximum in the summer of 1959 was 4·2°C and in 1960, 11·2°C), weak radiation and sparse plant growth. Vegetation has little effect on water flow: in fact, flow is furthered by the soil stripes running downslope. As a result, the transfer of rainwater and snowmelt water to the rivers is more rapid and complete than in any other climate.

Drainage flushing, solifluction and discrete ground-ice formation

The rivers in periglacial regions abrade the coarse rubble even more strongly than in the desert. This is not due solely to the supply of debris by solifluction (till now this has been somewhat over-estimated) but through the combined effect of three processes: drainage flushing (including mechanical eluviation), solifluction and discrete ground-ice formation. Beginning on the upper snow-free segments, slopewashing occurs in black stripes over the snow. With the advance of melting a thin film of water runs over the ground, which is as yet hardly touched by the thaw. As the thaw penetrates, so also does the meltwater. In the middle of the summer drainage flushing occurs at a depth of 30 to 40 cm, that is at the base of the active layer. Yet even here its vigour is not completely checked: for as the whole surface moves slowly down by solifluction continuously and the differential movements due to freeze–thaw have a loosening effect, there are always sufficient cavities within the regolith to allow water to flow at a depth of 30 to 40 cm. Quite often in summer the water can be heard rushing along beneath the coarse stripes in the thawing ground. This water is always cloudy with fines, never clear like trickling underground water.

In place of the purely surface washing of slopes in other climatic zones, *three-dimensional flushing* takes place here without any loss through seepage. This and solifluction stimulate each other. As soon as the slope exceeds 2°, the whole covering of rubble begins to move downslope by solifluction. More than 800 measurements taken on the slopes shown in Fig. 13.3, and made up of marly slate, arkose and basalt, yielded the following figures: coarse rubble stripes sloping at 6°, 0·6 cm per year, and on fine stripes 0·9 cm per year; on steeper slopes (12 to 18°) composed predominantly of fine material, movement of as much as 2·5 to 2·9 cm per year (in extreme cases 4·6 cm per year!) took place. Finally on an embankment of some 24°, 5·2 cm per year, with an extreme figure of 9·9 cm per year was recorded. The ground movement increases downhill while the incline remains the same, since the flow of water increases. Moreover, due to the wash from upslope there is a progressive downslope increase in the proportion of clay and silt. This powerful, three-dimensional downhill movement of the whole ground surface is unknown in all other climatic zones. Due to its constant mechanical movement it lacks the typical weathering shown by all other regolith on earth. With this transportation of immense surface rubble is associated an equally immense *erosion* on all types of slope mentioned. This is intensified by the ready incorporation of loosened debris from beneath (see Büdel, 1961a).

In the uppermost zone of the permafrost (that is directly below the 30 to 40 cm-thick active layer) there lies a *zone of frost-heaving* in which rock is completely broken up into rubble, through numerous joints and other planes filled with ice, including compact masses of segregated ground ice (Taber, 1930). It is in this zone that great variations in temperature and volume may be observed with an annual or seasonal rhythm. The regolith which glides over this plane (solifluction plus drainage flushing) does not excavate intact rock by mechanical corrasion; rather it picks up frost-heaved material because in warmer summers the active layer penetrates deeper so that entrainment of previously loosened material is effected *merely due to the melting process*. This is a very effective process and is particularly associated with tundra and frost rubble climates with perennially frozen ground. Under such conditions river erosion is stimulated. In summer the water level is high and there is a large volume of coarse debris which facilitates erosion. In the high polar regions of frost rubble (and the tundra zone in part) there are only small reserves of underground water due to restricted penetration of seasonal thaw. Accordingly, the rivers disappear completely after the first winter frost, because there are no sources lower down to feed them. Their beds, which are dry by early winter, freeze completely.

For this reason the permafrost and its upper, ice-heaved zone extends directly beneath the valley and its broad rubble floor. This is presumably the reason why such marked valley deepening is attained only in the tundra and frost-rubble zones and not in the subpolar forest region, although the permafrost penetrates in Siberia, Alaska and Canada far southwards into the forest belt. In this belt, the active layer is several metres thick and the larger rivers are fed from this reserve even in winter. For this reason, permafrost is not continuous and is often completely absent (as in parts of the Yukon and Alaska). This results in a reduction in linear erosion. It is for this reason above all that the equatorial limit of the zone of pronounced valley formation is placed at the polar forest limit in Fig. 13.1 and not at the equatorial limit of permafrost. The snowmelt water which fills the river bed anew early the next summer always has a temperature of 1 to 2°C and, due to its volume, it has a far greater heat capacity than the subsurface drainage on the slopes. Therefore, the melting of seasonal ice occurs much more quickly on the valley floors than on the slopes. The river effects little mechanical excavation by bed erosion: rather, due to melting on its bed, it incorporates frost-riven debris beneath and so removes it. In this way its debris load is further increased, and braiding and distributary development is thus stimulated. In the first instance, therefore, it is the effect of ground ice which gives the rivers of this climatic zone their ability to erode deeply and rapidly across broad valley floors, with simultaneous undercutting of the adjoining, gently concave valley-side slopes. This constitutes one of the main reasons for the special position which the 'zone of pronounced valley formation' occupies amongst the other climato-morphological zones. Another is the notable penetration of mechanical weathering into the permafrost, traces of which are known down to 350 metres in Spitzbergen. For reasons which cannot be discussed here (but notably the scarcity of terraces in the Spitzbergen valleys) the conclusion was reached that the valley-formation mechanism there is analogous to that deduced for Central Europe during the humid early parts of glacial periods rather than the pleniglacials.

Tropical zone of planation surface formation

The tropical zone of planation surface formation is the other extreme amongst the climatomorphological zones. The morphogenetic complex which characterises it is related (in its purest expression) to the climate of the seasonally humid tropics, with a clear dry period, and so to the different variants of the savanna climate with their tropical red earths rich in kaolin (Kubiëna, 1962, 1963). These soils form the upper organic layer of a widespread inorganic deep-weathering profile. The depth of weathering reaches some tens of metres on average on particularly old surfaces. In Indonesia (Banka Island for example) up to 170 m is attained, while in West Africa and India up to 300 m (and in extreme cases even up to 600 m) has been recorded (Tienhaus, 1963, von Gaertner, 1963). So like the perennially frozen ground, these areas of tropical weathering contain many characteristics which are still very active but have their origin in past geological periods. However, these are the product of purely *chemical* weathering. Even the uppermost horizons, which are characterised by almost complete mineral alteration to clay minerals commonly reach a depth of 10 to 30 m when maturely developed. Such weathering profiles are found on all gently sloping surfaces up to about 3° and often on gently rolling land up to an average slope of 8° to 12°. As in the polar regions, deep weathering favours certain combinations of exogenic processes. Let us consider the end-product, or the actual relief features of the tropical zone.

Rumpfflächen characteristics

In the middle latitudes extensive plains at river-level are only found in areas of deposition (Munich Plain, Marchfeld, Schütt Islands, Alföld, Plain of the Po, Wallachia). In the tropics, on the other hand, planation surfaces (*Rumpfflächen*) of gently rolling character extend right to the river courses, the maximum slope of the gulleys connecting these surfaces reaching only 3·5°. Such surfaces dominate those parts of the crust characterised by gentle to moderate epeiro-variance (intensity of uplift equal to or less than one cm per century). They are found in a particularly fine form in Guyana, the Deccan and above all throughout tropical Africa. The best example of this type is the Inner Sudan *Rumpfflächen* which stretches from the edge of the Sahara in the north to the Upper Guinea and Asande threshold in the south, about 500 km wide and more than 5000 km long from Senegal to the Nile right across the continent. This surface may be replaced by similar but higher surfaces from time to time. These are often 500 or even 1000 m above sea-level, but they have the same features: characteristic *Rumpfflächen*, undissected at river-level. Nothing like this is possible outside the tropics where all such surfaces (French Central Plateau, Rhineland and Schiefergebirge) have long ago lost this character and have developed deep valleys.

Inselberg characteristics

These surfaces are interrupted only by inselbergs which generally rise steeply above them. Inselbergs are due only partly to the presence of resistant bedrock. The fringing mountains, like the inselbergs, rise sharply above the plain to the next highest surface. The outlying

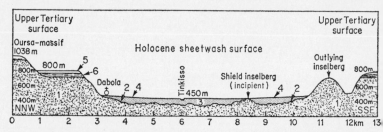

Fig. 13.4   The sheetwash surface on the Tinkisso—source stream of the Niger near Dabola (Guinea)—as an example of recent erosion surface development

1—dolerite bedrock; 2—surface at the base of the zone of weathering; 3—10 to 30 m thickness of kaolinised red loam; 4—subaerial sheetwash surface; 5—lateritic iron crust, 2 to 4 m thick; 6—bauxitised zone beneath (5), 10 to 15 m thick. (5 and 6 are elevated remnants of an older surface.)

inselbergs often display remnants of these higher surfaces. This is a sign that the plain is continually growing laterally, the sharp break of slope at the hillfoot being maintained. Apart from these 'outlying inselbergs' there are also the 'azonal' (Kayser, 1949) or 'shield inselbergs' which are considerably smaller subsurface knolls, often distributed indiscriminately across these surfaces.

The mode of formation of such surfaces has been dealt with in detail elsewhere (Büdel, 1957a, 1957b). It is based on the concept of 'double planation surfaces' (Fig. 13.4). The two surfaces are the upper and lower boundaries of the weathering profile. The basal surface or weathering front is irregular, penetrating well-jointed zones and avoiding areas of particularly resistant rock which may well appear on the surface as shield inselbergs. In such a case they emerge into an environment which is only seasonally moist (it is desert-like in the dry season); this is completely different from the ubsurface environment which is continually moist and hot and favours chemical decomposition. Weathering takes place on this lower levelling surface *only*—there is no erosion or mechanical removal of the weathered mass. Thus the material is a purely local weathering residue (autochthonous), commonly displaying an old, polygenetically matured profile (Kubiëna, 1962, 1963).

Erosion, on the other hand, is restricted to the upper levelling surface. Even the largest of the rivers flow in the dry period in flat channels only a few metres below the general surface. In the rainy season they rise quickly and, with their larger tributaries and numerous shallow flood channels, inundate the area to produce the whole gently rolling relief of these surfaces. (On the gentle divides between such wide floodwater channels, inselbergs of both types tend to be situated; quite often they are indistinguishable lithologically from the surrounding surface.) In this way the whole surface is planed to match the gentle longitudinal gradient of all the drainage channels. Large parts of this surface remain undissected (until the slope exceeds 3·5°). The rivers cut into the deeply weathered zone even in the low-water period; they seldom affect sound rock and lack the weapons of erosion to do so. As Bakker (1957) established in Surinam, the fluvial sediments are of the same mixture of fine sand and clay as the ground the river flows over. Where the rivers cross resistant rock ridges (such as the shield inselbergs), they fail to incise them, and instead rush over in waterfalls or rapids (Sulas, Cachoeiras). This is the cause of the many cataracts typical of tropical rivers (Nile, Niger, Congo, Amazon tributaries). The longitudinal gradient between waterfalls is minimal. Tropical rivers, even far inland, lack the regular, longitudinal profile of extratropical rivers, particularly those which flow in valleys which were shaped (today or previously) under a permafrost climate.

It is also very characteristic that inselbergs and foothills rise above such surfaces usually with relatively steep, often smooth slopes, mostly lacking re-entrants. Gentle and steep surfaces meet abruptly: there is no transition from surface to slope, gully, gentle foothill and valley, such as may be seen in the Mittelgebirge of Europe. There too, of course, the higher mountains are dissected. Slightly hilly areas with a relative relief of 50 to 100 m within distances of 2 to 5 km are seldom found in the tropics, although they sometimes occur, for example in parts of Somaliland, Upper Volta and the 'coastal plain' of Ghana. In the case of the latter, valley incision is beginning on a wide, flatter and older coastal plain, possibly owing its origin to the eustatic retreat of the sea during Pleistocene time. In such an exceptional case the use of Louis's term 'gulley-valley' is appropriate.

Uplifted tropical surfaces

High old surfaces occur often enough both outside and within the tropics, for example the Altiplano in eastern Bolivia, the Amhar Highlands of north Ethiopia, and, in Europe, high surfaces which extend from the Jotunheim via the Harz to Dachstein. In the tropics, however, it is only in exceptional cases that they are truly relict 'old surfaces', so important is the role of tectonic uplift. These uplifted tropical surfaces are mostly covered by red loam and substantially uncut by valleys. Above all they show the same tendency to *lateral expansion* at the expense of the surrounding mountain areas already shown to be typical of lowland surfaces. On such high surfaces there are often shallow valleys which form steep gullies as they descend to the next flat surface and again to the surrounding major lowland surface. The eastern flank of the Amhar Highlands and the northern edge of the Fouta-Djalon mountains show this. This is not the case *anywhere* outside the tropics and W. Penck was clearly mistaken about this. Even where the main valleys of markedly uplifted tropical mountains have cut right through all such surfaces due to their large discharge and load (as in western Ethiopia and eastern Bolivia), in cross-section they are much more rugged and narrow. Above all, their longitudinal profiles are much steeper (containing abundant breaks of slope and waterfalls) than comparable examples from extratropical regions. Particularly noticeable steps appear in Ethiopia at the mouths of smaller tributaries. The mountain valleys of the tropics lack the smooth longitudinal profiles and to a great extent also the broad, level valley floors of the extratropical valleys. Such are found only in the Himalayan and many Andean valleys, particularly in the upper reaches, where the influence of high altitude brings into operation 'extratropical' or even 'pronounced valley formation' conditions.

DIFFERENCES BETWEEN CLIMATIC AND CLIMATOGENETIC GEOMORPHOLOGY

The last part of this article deals with the difference between 'climatic' and 'climatogenetic' geomorphology and is particularly concerned to demonstrate the necessity of progressing from the first to the second. Again I take the extratropical regions (essentially the middle-latitudes) as a basis.

Although in middle-latitudes the attainment of a climax or maturity of development at a single point (for example of a soil profile on loess or in the regular longitudinal profile of a small valley) may possibly occur 'quickly' (in 10^4 to 10^5 years), it takes a very long time (on average for all climatomorphological zones of the earth some 10^7 years) for a whole landscape—from base-level to watershed—to be rendered in harmony with conditions prevailing in any one of the *present* climatomorphological zones. During such long periods of time, however, there often occur changes in climate, that is another combination of processes becomes dominant annulling earlier relief features, stamping them as 'dead' forms and working towards a new relief to match the new combination of processes; this is termed a *relief-climax*.

In the middle-latitudes such changes have occurred several times within the last 10 to 15 million years; these changes encompass almost all of the climatic variations from the edge of the tropics to the Pole (see Fig. 13.2). The result is that everywhere on earth and particularly in the middle-latitudes—apart from the two extreme zones of the present tropical and polar regions which form relief features quickly and characteristically—in addition to the forms arising from current climatomorphological process complexes, there also exist forms due to former complexes (called 'ancient forms' in the earlier literature, for example by Passarge, 1919 and Mortensen, 1929). Where several changes in climate have occurred within the last 10^7 years it is possible, therefore, to *distinguish several relief-generations* in the present-day relief. It is the task of climatogenetic geomorphology to make this distinction as exactly as possible.

Isolated single processes in geomorphology

Accordingly, a three-tier relationship may be discerned in modern geomorphology. The first is the study of *isolated single processes*, for example the weathering of rock masses (Wilhelmy, 1958), the various soil types (Kubiëna, 1962, 1963), the formation of river courses (Hjulström, 1935: Strahler, 1956) and so on. Like Tricart (since 1951) one could call this approach 'dynamic geomorphology'. Since, as we have seen, each of these processes has distinctive characteristics and a different sequence and culmination in different climatic zones, this dynamic part of modern geomorphology cannot be rigorously pursued without reference to the climatic zones.

The second level, 'climatic geomorphology' deals with the study of climatic variation, that is, it is not concerned merely with single processes but rather with all currently active geomorphological processes; it tries to distinguish the particular results of their combined activities as they vary with climate, and so to distinguish the present-day climatomorphological zones. The so-called aclimatic factors are expressed differently in the relief from one climatomorphological zone to the next.

Analysis of relief

Climatogenetic morphology, then, is the *analysis of the relief itself* and is somewhat like a photograph in its completeness. First, it tries to isolate the elements of the relief due to current processes (presuming the tasks posed at the second level have been solved). It then deals above all with the recognition of the various older generations of landscape produced by fossil climates. Little work of this kind has been carried out: in my view it is *the central task of geomorphology*. The few observations at my disposal, in the longitudinal zone through Central Europe from Spitzbergen to the lands of the Sudan, suggest these three conclusions.

(1) Here (and elsewhere on earth) from the Pole to the Equator several landscape generations are discernible.

(2) Everywhere these relict landscapes form a far greater part of the present land surface than the relief elements formed by present-day processes.

(3) For this reason present-day processes cannot operate freely but are forced into certain narrow courses by the relief elements already in existence. (These influences can never be taken into consideration fully by a laboratory experiment, hence the limited use of this technique in dynamic geomorphology.) That this applies particularly to the middle-latitudes is shown in the ideal profile through a river valley in the German Central Uplands (Fig. 13.5). Five major morphological elements can be distinguished. The first element is the principal old surface; on the River Main this is the approximately 300 metre-high Gau surface. Although inherited from Tertiary predecessors its latest expression is found in the upper

Pliocene. Its age is proved by old remains of red loam found in karstic depressions and other crevices, on the few older Franconian Alp surfaces and on the Rhône in numerous fossil finds. Its origin is similar to that of the tropical zones of planation surface formation. The presence of flat knolls scattered indiscriminately over these surfaces, like the tropical shield-inselbergs, agrees with this interpretation (Büdel, 1957c). These relief-generations of Pliocene age remain the dominant element in the present Franconian landscape.

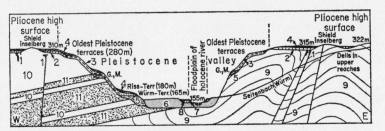

Fig. 13.5 The relief-generations of Central Europe, expressed in an idealised profile of one of the German Mittelgebirge valleys

First relief-generation: Pliocene surface, undistorted by rock variations, watershed migration and elimination; 1—karstic pipes and weathering joints with Pliocene red or brown loams; second relief-generation: oldest Pleistocene terraces; third relief-generation: Pleistocene valley; 2—cryoturbate mantle of the old surface; 3—solifluction mantle of the slopes; 4—loess mantling 2 and 3; 5—*schotter* gravels of the old and middle Pleistocene terraces (G, M = Gunz, Mindel respectively); 6—*schotter* of the upper Würm terrace (pleniglacial); fourth relief-generation: late-glacial and Holocene prior to the Neolithic; 7—lower Würm terrace (late-glacial); fifth relief-generation: post-Neolithic Holocene; 8—loess mantle on 7; floodplain of present river—subsurface data; 9—variable structures (folded crystalline rocks); 10—morphologically resistant mezozoic beds; 11—morphologically weak Mezozoic strata

In comparison, the second relief-generation is less extensive. It originated over a relatively short period, bracketed at beginning and end by two radical changes of climate; this is the oldest Pleistocene stage (Villafranchian, see Fig. 13.2 and Büdel, 1963(*b*)). For the first time since the long period of surface wash of the 'old tropical earth' (Eocene to the Miocene–Pliocene transition) the rivers acquired the power to deepen in very wide but still rather flat valleys (Büdel, 1957c). On the Rhine and other rivers they are known as 'trough surfaces', while on the Main they were at first called 'transition terraces' by Körber (1963) and myself and later 'earliest Pleistocene terraces'. Compared to the preceding Pliocene period, the climate was colder, but for a time drier. The rivers received coarser tools of erosion. The outlines of the present-day river systems and watersheds were determined; the Danube lost the upper Main and the Regnitz trench to the Rhine system.

The third relief-generation comprises the narrow Pleistocene valley trenches which developed in the change from warm to cold periods, chiefly under tundra and frost rubble climates, from Sizil and pre-Gunz to Würm; these valley trenches are found everywhere on the old surfaces of Central Europe. Their development began directly after the Villafranchian with the deterioration of the climate into the first cold period and worked its way step-by-step to the present-day levels, which are on average 180 to 250 m lower (in exceptional cases even 300 m) than the adjacent surfaces. The creation of the whole concave form (the 'Pleistocene Valley', in Fig. 13.5) with its rounded and smoothed slopes, is the work of the Pleistocene period as are the terraces and also the present valley floors which are, in fact, to a large extent identical to the Würm terraces. The cryoturbate mantles date mostly from this time, and cover the remains of Pliocene debris on the old surfaces. They are also found on the more or less uniform terrace surfaces, on the Pleistocene valley slopes, or as solifluction blankets on all steeper slopes. In the most continental climatic region of Central Europe, loess covers are often present (apart from the Würm terraces which are always free from loess). In the Würm, most of the small lateral valleys received their final shaping, in particular their now-dry source areas and their floors which run down to the Würm terrace, not to the level of the present-day river.

The fourth relief-generation is of Holocene (prehistoric) age, being in fact, largely late-glacial. Since the work of Mensching (1950) a narrow zone on either side of the present river below the lowermost terrace has been allocated to the late-glacial period. As a general rule the larger the present river the larger is the proportion of the valley bottom this zone occupies (Bremer, 1959).

The upper floodplain loam sheets, and the small lateral gorges which cut through the valley bottoms to present river levels are the product of the fifth (Holocene-historical) relief-generation: that due to human influence.

Of course, it is not only in middle-latitudes that the dominant process–complexes of the Holocene period were so relatively ineffective that earlier relief-generations now occupy a larger surface area than the recent ones. Everywhere in the world the high-lying trunk surfaces ('old surfaces'), which date from the uniform warm climate of the Eocene to Mio-Pliocene period ('old tropical earth') are a major morphological component. In the zone of pronounced valley formation such surfaces are better preserved than had previously been assumed (Fig. 13.3, see in particular Wirthmann, 1962). They also play an important role in the inner tropics such that higher trunk surfaces, despite having been formed in prehistoric times, may still preserve certain characteristics of active surface formation. In the few tropical areas of the earth which have not experienced a morphologically effective climatic change from the early Tertiary to the present, there appear relief-generations of tectonic rather than climatic type. Over the largest part of the earth, however, the visible relief-generations have been created climatogenetically rather than tectonically.

In all other climatic zones, however, from the tropical, through the major deserts, the subtropical areas of winter rain, the extratropical steppe to the forest areas and even to the Polar zones up to the edge of the large inland ice sheets (even *under* these), and even under recent marine mantles on continental shelves (like that of the Barents Sea of Wirthmann, 1962)—everywhere there appear to exist crowning, dominant relief elements such as 'old surfaces' derived from the period of 'old tropical earth' formation. This often occurs in such a pronounced way, that the task of climatogenetic geomorphology is to a large extent the recognition and distinction of the type, degree and stages in the destruction and decomposition of these old surfaces.

It should now be clear that climatogenetic geomorphology is the third step (dynamic and climatic geomorphology being the first and second) which systematically widens and supplements the framework of geomorphology as it has been practised hitherto.

REFERENCES

BAKKER, J. P. (1957). Quelques aspects du problème des sediments correlatives en climat tropical humide. *Ztschr. f. Geomorph.* N. F. Bd. 1

BOBEK, H. (1953). H. Lautensachs geogr. Formenwandel—ein Weg zur Landschaftssystematik. *Erdkunde* 8

BREMER, H. (1959). Flusserosion an der oberen Weser. *Göttinger Geogr. Abh.*, H.22

BREMER, H. (1962). Morphologische Spuren des pleistozänen Klimawandels in Zentralaustralien. Vortrag DEUQUA–Tagung, Nürnberg

BUDEL, J. (1935). Die Rumpftreppe des westlichen Erzgebirges. *Verhdl. Dt. Geographentag*, 25 Bd .

—— (1948). Das System der Klimatischen Geomorphologie. *Verhandl. Deutscher Geographentag*, 27, 65–100

—— (1957a). Die doppelten Einebnungsflächen in den feuchten Tropen. *Ztschr. f. Geomorph.* N. F. Bd. 1

—— (1957b). Die Flächenbildung in den feuchten Tropen. *Verhdl. Dt. Geographentag*, 31 Bd

—— (1957c). Grundzüge der klima-morphologischen Entwicklung Frankens. *Würzburger Georg. Arb.* H. 4/5

—— (1961a). Die Abtragungs-Vorgänge auf Spitzbergen im Umkreis der Barentsinsel auf Grund der Stauferland-Expedition 1959/60. *Verhdl. Dt. Geographentag*, 33, Bd.

—— (1961b) Morphogenese des Festlandes in Abhängigkeit von den Klimazonen. *Die Naturwissensch.* 48, H. 9

—— (1963). Die pliozänen und quartären Pluvialzeiten der Sahara. *Eiszeitalt. u. Genenw.* 14

CORBEL, J. (1957). Les Karsts du Nord-Ouest de l'Europe et de quelques régions de comparaison. *Inst. etud. rhodan. Univ. de Lyon, Mem. et Doc.* 12

DAVIS, W. M. (1912). *Die erklärende Beschreibung der Landformen*, Leipzig

GAERTNER, H. R. VON (1963). Tropische Tiefenverwitterung in den Nimba-Bergen (Grenze Elfenbeinkuste, Guinea, Liberia)

HJULSTRÖM, F. (1935). Studies of the morphological activity of rivers as illustrated by the River Fyris. *Univ. Upsala Geol. Inst. Bull.* 25, 221–527.

KAYSER, K. (1949). Die morphologischen Untersuchungen an der grosslen Randstufe an der Ostseite Sudafrikas. In: Obst und Kayser. *Die gr. Randst. a. d. O-Kuste S-Afrikas.* Hanover

KNETSCH, G. (1959). Uber aride Verwitterung unter besonderer Beruksichtigung naturlicher und kindstlicher Wande. *Ztschr. f. Geomorph.* N. F. Suppl.-Bd. 1

KÖRBER, H. (1962). Die Entwicklung des Maintals. *Würzburger Geog. Arb.* H.10

KUBIENA, W. L. (1962). Polygenetische Bodenkunde und Aufbauelemente der Tropenböden, *Vortrag Univ.* Hamburg am 12. 12

—— (1963). Die Genese lateritischer Profile als bodenkundliches Problem. *Vortrag v. d. Ges. Dt. Metallhütten-Bergleute*, Würzburg am 1. 3

LAUTENSACHS, H. (1953). Der Geographische Formenwandel. *Coll. Geogr.* 3. (Bonn)

LEHMANN, H. (1955). Der tropische Kegelkarst in Westindien. *Verhdl. Dt. Geographentag.* Bd. 29

MENSCHING, H. (1950). Schotterfluren und Talauen im Niedersächsischen Bergland. *Göttinger Geogr. Abh.* H. 4

MORTENSEN, H. (1929). Über Vorzeitformen und einige andere Fragen in der nordchilenischen Wüste. *Mitt. Geogr. Ges.* Hamburg

—— (1958). Über Wandverwitterung und Hangabtragung in semiariden und vollariden Gebieten. *Congr. Int. de Geogr.* Rio de Janeiro (1956).

PASSARGE, S. (1919). Die Vorzeitformen der deutschen Mittelgebirgslandschaft. *Petermanns Mitt.*

PENCK, A. (1910). Versuch einer Klimaklassifikation auf physiographischer Grundlage. *Sitz.-Ber. Preuss. Akad. d. Wiss. Phys.-Math.* Kl. 12

PENCK, W. (1924). *Die morphologische Analyse*, Stuttgart.

STRAHLER, A. N. (1956). 'The nature of induced erosion and aggradation', in:

Thomas, W. L. Jr. (Ed.) *Man's Role in Changing the Face of the Earth*, University of Chicago Press

TABER, S. (1930). The mechanics of frost-heaving. *J. Geol.*, **38**, 303–17

TIENHAUS, R. (1963). Verwitterungsprofile uber Itabiriten von Afrika und Indien. *Vortrag v. d. Ges. Dt. Metallbutten-Berglente*, Würzburg am 1. 3

TRICART, J. (1951). Le modelè periglaciaire. Cours de géomorphologie, 2ᵉ: partie, fasc. 1., Paris, C.D.U.

WILHELMY, H. (1958). *Klimamorphologie der Massengesteine*. Braunschweig

WIRTHMANN, A. (1962). Die Landformen der Edge-Insel in Sudostspitzbergen. *Würzburger Naturw. Habil.-Schrift 1962, erscheint in den Ergebnisberichten der Stauferland-Expedition*

14 Morphogenic Systems and Morphoclimatic Regions

JEAN TRICART and ANDRÉ CAILLEUX

Within any one morphogenic system, the various morphogenic processes are not all of the same importance. Some play a role which is both very general and very intensive, for example the action of variations in temperature and humidity on rocks in arid regions, or freeze–thaw action in periglacial environments. When such processes are so intensive and widespread that they largely determine the landform characteristics of the morphogenic system they may be referred to as *dominant processes*. To these may be added the *subsidiary processes* which are more limited in their action, in that they affect certain rocks only, or act only from time to time.

This distinction between dominant and subsidiary processes is very important in geomorphology, and it will undoubtedly be further justified by increasing numerical work as the discipline progresses and so form a basis for the morphoclimatic subdivision of the globe.

DOMINANT AND SUBSIDIARY PROCESSES

Frost-shattering is one of the most typical processes which depend directly on the climate. It depends first of all on oscillation of the temperature above and below 0°C or on the number of annual freeze–thaw cycles. Secondly, it is influenced by the length and severity of such cycles, as they control the depth of penetration of cold waves into the ground. As these vary so is there a spatial variation in the associated phenomena and their geomorphic results.

For example, in ice-cap regions (Greenland, the Canadian Arctic archipelago, Spitzbergen, the Siberian Islands and Antarctica) freeze–thaw cycles are small in number. The summer thaw is brief and lasts for only a few weeks, nocturnal frosts being frequent so that frost-shattering is important at this time. The total number of freeze–thaw cycles, however, remains generally less than forty. Limited periods of observation lasting one to two years have shown that in many places in Antarctica the air temperature does not rise

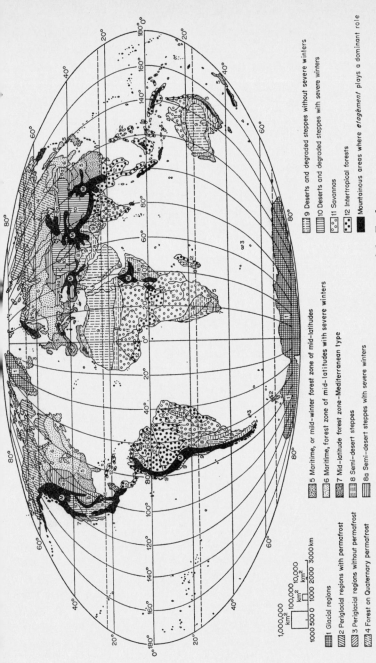

Fig. 14.1 Morphological–climatic zones of the Earth

5 Maritime, or mild–winter forest zone of mid-latitudes

6 Maritime, forest zone of mid-latitudes with severe winters

7 Mid-latitude forest zone–Mediterranean type

8 Semi-desert steppes

8a Semi-desert steppes with severe winters

9 Deserts and degraded steppes without severe winters

10 Deserts and degraded steppes with severe winters

11 Savannas

12 Intertropical forests

12 Mountainous areas where *étagement* plays a dominant role

1,000,000 km² 100,000 km² 10,000 km²

1000 500 0 1000 2000 3000 km

1 Glacial regions

2 Periglacial regions with permafrost

3 Periglacial regions without permafrost

4 Forest on Quaternary permafrost

above freezing-point. It may be concluded from this that the nuna-taks rising above the polar ice sheets are not greatly affected by frost-shattering. This would also seem to explain the great scarcity of supraglacial moraines on these ice sheets, even adjacent to nunataks. By way of contrast, in the Eqe area (N.E. Greenland) and further south, frost-shattering is very important and the land-scape is rapidly modified due to the rapid fragmentation even of the most resistant rocks.

The number of freeze–thaw cycles is greatest at sea-level in the circumpolar rock deserts where the mean annual temperature is approximately 0°C. In the maritime areas of this zone, for example Iceland or the Kerguelen Islands, freeze–thaw cycles exceed 150 per annum. This is essentially a daily régime with each cycle being very short. As a result cold waves penetrate only a few centimetres into the ground. Frost-shattering is then severe but superficial and is a function, in particular, of rock microstructure and porosity; it may be termed *micro frost-shattering*. In the more continental areas, however, such as Northern Canada or Siberia, the number of freeze–thaw cycles is smaller but the temperature range greater, the cycle tending toward a seasonal rhythm. In particular, it is during the change of season, in spring and autumn, that freeze-thaw cycles are most frequent. The number of freeze–thaw cycles in such regions is generally between 20 and 50 per annum. A further point of differ-ence is that the summer thaw penetrates deeply below the ground surface, the active layer affected by such fluctuations being 1 to 2 m deep. Thus, freezing is able to exploit weaknesses in the rock and loosen joint-bound blocks; this is termed *macro frost-shattering*.

Beyond the rock deserts a new factor is introduced in the form of vegetation. Frost-shattering and a vegetal cover are diametrically opposed. If frost-shattering is sufficiently great to inhibit the development of vegetation, plants commonly colonise only those areas unaffected by geliturbation, such as sands and gravel, leaving the more susceptible clays and silts bare. Vegetation is important in that it can delay the penetration of cold waves. This is the result not only of the presence of the vegetation itself but of the existence of a peaty layer. In vegetated areas, therefore, freeze–thaw cycles of intensity and frequency equal to those found in the tundra and rock desert regions may have a much less marked effect.

This relationship is also important in the forested regions. The Taiga with a soft, peaty ground surface, makes a very good insulator.

Perennially frozen ground formed during the last Pleistocene glaciation is perpetuated by the differential absorption of solar energy by the vegetal cover and is thus preserved as a relic feature. As has been shown in Alaska, clearing of the woodland results in the development of a chaotic microrelief resulting from the differential melting of the perennially frozen ground arising from increased ease of penetration of warmth.

Vegetation can be regarded as a dominant geomorphic factor in the temperate zone. Soils are often thick enough to prevent cold waves from reaching bedrock. In Alsace, for example, where an average of 30 freeze–thaw cycles occurs each year with temperatures dropping to −15° or −20°C, frost-shattering is almost entirely negligible. Penetration of frost in the most severe winters is never more than 60 cm. By way of contrast, frost-shattering is more effective in the *garrigues* of Languedoc, well to the south of Alsace, where minimum temperatures rarely fall below −5° or−7°C and where the freeze–thaw cycles are much fewer. This apparent paradox is explained by the fact that humanly induced modification of the vegetation has left the surface bare, so that it is susceptible to severe frost action of the micro type. In snowy regions, low total precipitation and wind action are additional factors which favour frost action in that they effectively reduce the insulation which a thick snow cover provides.

The example of the varying importance of frost-shattering illustrates the principle of morphoclimatic zonation as follows.

(1) In the polar periglacial zone, frost-shattering is the dominant process which, in association with geliturbation, renders the periglacial morphogenic system distinctive. Its effects influence the development of all landforms.

(2) In glacial areas of the polar regions, where temperatures are almost always below the freezing-point, frost-shattering is a subsidiary process only, the dominant process being moving ice in the form of glaciers.

(3) In parts of the boreal conifer zone, frost-shattering, together with geliturbation, remains a dominant morphogenic process mainly due to the persistence of deep permafrost.

(4) In maritime areas of the mid-latitude mixed-forest zone, frost-shattering is no more than a subsidiary process; its importance increases toward the continental interiors and the cold steppes with a sparser plant cover.

If we extrapolate from the morphogenetic process to the more general level of the morphogenic system, it may be noted that when the system is dominated by physical processes, the biochemical processes are subsidiary, and vice versa.

MORPHOGENIC SYSTEMS IN WHICH PHYSICAL OR BIOCHEMICAL FACTORS ARE DOMINANT

The example of frost-shattering serves to demonstrate that climatic geomorphology does not simply mirror climate. This specific example serves to illustrate the basic distinction between the physical world and the biotic world. Accordingly, we may distinguish two major categories of morphogenic system: one in which physical processes are dominant and a second in which biotic processes dominate.

Domination by physical processes

Morphogenic systems dominated by physical processes are found in those areas with a sparse and discontinuous plant cover, that is cold or arid regions of both high and low altitudes. Generally speaking, soils are thin and develop very slowly, so that their insulating effect is rather limited. As the effects of climate are felt directly in such conditions, the morphoclimatic systems are ones in which mechanical processes typically dominate the chemical processes and, to an even greater extent, the biochemical processes. The limits of such regions, then, are ecologically defined; they correspond to the margins of a plant cover which is sufficiently dense to affect the morphogenic processes.

Of course, boundaries such as these are rarely sharp ones. Thus the tundra regions form a transitional zone between the Arctic rock deserts and the boreal conifer zone. In the tundra the opposed interplay between the mechanical and biotic processes is especially evident, as areas without vegetation occur because of snowdrifting and the mechanical disturbance due to geliturbation. Elsewhere, the mechanical processes are partly inhibited by the presence of tundra vegetation or grass so that modified forms of solifluction operate beneath the vegetation mantle. A transitional zone is also found in the steppe regions on the margins of the arid zone in middle as well as low latitudes. Here the sparse vegetation only marginally modifies the effect of runoff and wind action. Wind is just as important and severe showers with alternation of wet and dry periods probably more important than they are in adjacent more arid areas with less complete plant covers. There is a particularly delicate balance between the vegetation and the mechanical processes in such areas, and such transitional zones are particularly susceptible to variations in morphogenic conditions. Climatic oscillations or man's activities may have results which appear inordinately severe. We have already noted the rapid changes which may accompany the clearing of the tundra woodlands. The reason for this is that the plants of this zone exist at their ecological limits; their sudden removal in effect eliminates the area's only defence against powerful mechanical processes.

It follows from this that the boundaries of morphoclimatic regions dominated by physical processes are determined by an ecological complex arising from several factors: weather and climate, type of soils, including paleosols, and vegetation cover. In contrast to the views of Davis, relief development is not dependent on a single process in areas with a morphogenic system dominated by physical processes. Desert morphology is as little dependent on the sole action of the wind as is glacial morphology on the single process of ice abrasion.

Domination by biotic and pedological processes

Morphoclimatic regions dominated by biotic and pedological processes characteristically display soils which are derived from *in situ* bedrock modification. Biotic and chemical processes are the only ones able to maintain such soils in the face of the mechanical processes which tend to destroy them.

In humid regions erosion is mainly by way of biotic or chemical processes and the bulk of the material transported to the oceans in solution exceeds that which is carried in suspension. Plant cover serves to inhibit mechanical activity although this is evident in the top soil; sheet erosion is quite limited under a forest cover and is even more restricted on dense grass covers. Creep is not very effective and burrowing and the uprooting of trees by the wind and landslides are often the most active processes affecting hillsides. The mechanical action of flowing water affects only the stream bed. Wind only rarely affects the soil.

In those morphogenic systems dominated by biotic factors, several types may be recognised on the basis of the degree of this dominance. Biotic factors appear to be strongly dominant in the tropical forests.

The plant cover in these areas is dense and breaks in the canopy ar
few and far between. Rock outcrops are exceptional even on the bank
of rivers. Beneath the tree canopy, overland flow has only a ver
limited effect and the sedimentation on valley floors is very slow
Rarity of outcrops, the downhill movement of debris other than cla
and the rapidity of chemical attack all tend to produce a dominantl
solutional load in the streams. The most important mechanica
process is that associated with landslides, especially on steep slope
as a result of the loss of dissolved matter, the uprooting of trees an
the undercutting of river banks.

In the temperate mixed forest zone, biotic and chemical processe
are generally less active in comparison. Here the vegetation undergoe
a season of dormancy, which may be a cool season or a dry seaso
(as in the Mediterranean), which serves to slow down the pedogeni
processes. As the regolith is thinner than in the humid tropics
bedrock is more readily attacked by the mechanical processes
Moreover, lower average temperatures reduce the rate of chemica
alteration. The higher proportion of clastic load carried in middle
latitude streams is the main mechanical erosional agent in this climati
environment; it influences the mechanical processes of slope development including rock fall, solifluction, landslide, earth flow and creep
Stream incision instigates the most active slope-forming processes
for overland flow is just as unimportant in this environment as it is
in the tropical forest. If the tree cover in middle-latitudes makes up
a less complete canopy to protect the soil than in the tropics the
ground surface is covered with a much thicker litter made up o
dead leaves and other decomposing debris such as mosses, lichen
and fungi which absorb water like a sponge.

The savannas, too, have their own distinctive balance between the
biochemical and the mechanical processes. The more variable plant
cover results in quite major variations in the moisture content o
the soils. The combination of a dry season and a herbaceous cover
is conducive to fires, some of which are spontaneous. The silica
content of the grasses is thus liberated and later removed by mechanical means, either by overland flow, wind or in solution (this is an
amorphous hydrated silica—opal—ten to twenty times more soluble
than quartz). The general impermeability and low water-holding
capacity of indurated (cuirassed) soils also favour runoff so that
the latter plays a more important role than it does in the rain-forest.
Streams undermine the margins of the duricrusts and so help to

dismantle them, particularly by removing the friable materials beneath them.

As one moves from the savannas to the steppes, so the mechanical processes in the morphogenic system become more important; there is progressively less corrosion and more corrasion by water, so that the streams are much more like those of the humid temperate regions.

MORPHOCLIMATIC REGIONAL BOUNDARIES

Boundaries of morphoclimatic regions may coincide with certain climatic parameters. For example, the poleward limit of the boreal conifer zone corresponds roughly with the 10·5°C isotherm for the warmest month. Generally speaking, however, it is a combination of factors rather than a single climatic factor which determines such boundaries and this is more in line with the concept of an ecological climate. For our present purposes we shall follow Birot who defines a dry month as one with precipitation in millimetres less than four times the average monthly temperature ($P < 4T$). The use of such a formula is no less artificial a procedure than the utilisation of other formulas that have been worked out in a more sophisticated way and, generally speaking, it provides satisfactory results. From a scrutiny of all such attempts the following conclusions may be drawn.

(1) The climatic factors which delimit the main types of world vegetation must be treated as a complex. Temperature and precipitation must be treated together, for example, in order to define the concept of drought.

(2) Certain major factors are present and these must be used to delimit the major bioclimatic zones, for example the continuous high temperatures of the tropical lowlands, the deficiency of heat in the cold zone, and the lack of precipitation in the arid zone.

(3) Variations and annual extremes in both temperature and moisture supply are also important. For example, according to Fournier (1960), the main factor affecting the intensity of erosion by surface runoff is an irregular precipitation régime.

THE ZONAL CONCEPT

While geomorphology under the influence of W. M. Davis was using an abstract theoretical method, other branches of physical geography saw the development of the zonal concept.

The concept has been defined mainly by Dokuchaev and, thereafter, Berg and Grigoriev, from research in pedology, biogeography and climatology. Thus, some branches of physical geography have had more than half a century's experience in handling the zonal concept than has geomorphology. It is useful to benefit from their experience and, in particular, to borrow some of their terminology especially as this appears to be well adapted to climatic geomorphology. Troll (1944) has already done this in his work on periglacial micromorphology.

Every characteristic or process with a distribution which broadly conforms to the latitudinal belts is termed *zonal*. The tropical rain forests, coral reefs, and the polar ice sheets are all zonal phenomena. A geomorphological assemblage which derives its character from processes typical of a latitudinal zone is a zonal phenomenon also, and may be termed a *morphoclimatic zone*.

The adjective *azonal* describes processes or phenomena which occur in several zones. For example, coastal wave action is not associated with any particular climatic zone. Vulcanism and mountain building are also azonal phenomena, even though certain orogenic belts, such as the Tethyan zone, have an east-west disposition.

If a phenomenon is characteristic of a certain zone but, nevertheless, occurs in another zone, perhaps in a more limited, sporadic or less well-developed way, it may be termed *extrazonal*. Thus Troll has described glaciers on equatorial mountains and their associated periglacial phenomena as extrazonal. Another excellent example is provided by the savannas which occur as clearings within the tropical rain-forests, as is the case in the Lower Ivory Coast or inland of the coastal strip in the Guianas. The adjective *polyzonal* or *plurizonal* may be used with reference to phenomena or processes which are not worldwide in distribution but which are found in several zones. A good example is running water which occurs mainly in the temperate and intertropical zones but which plays only a subsidiary role in the glacial and arid zones. The landforms which develop under forest are typically polyzonal, being found in the tropical rain-forests as well as in the mid-latitude forests.

THE RELATIONSHIPS BETWEEN ZONAL, AZONAL AND EXTRAZONAL MORPHOGENIC PROCESSES

Structural geomorphology provides the most typical examples of azonal phenomena. Volcanoes are perhaps the best example, their distribution being in no way zonal, cutting across the various morphoclimatic zones. Vulcanology and zonality come together only when we study erosion forms so that as the destructive forces of erosion slowly predominate over the constructive forces, the zonal elements become progressively more important. Thus the relationship between zonal and azonal characteristics varies with time.

Tectonic features appear to be less obviously azonal. The effect of the earth's rotation on the localisation of the major structural components of the earth has not yet been demonstrated but it cannot be excluded. In contrast there is less doubt about the importance of sedimentation, which is partly controlled by zonal factors, upon the development of fold mountains and isostatic movements for example. It may be that the rapidity of landform development in semi-arid continental climates accelerates tectonic development with the assistance of isostatic readjustments, the rapid erosion of raised blocks accelerating their uplift and the equally rapid infilling of trenches increasing their rate of subsidence. Indeed, it is in morphologically active semi-arid continental areas that the best examples of regenerated residual geosynclines have been found, in for example the Western United States, Central Asia and north-west Argentina.

In the realm of exogenic forces, the azonal factors are clearly evident. Submarine relief, coastal processes, wind action and running water are all classified in the azonal group. The mechanism of clastic sedimentation is essentially azonal, for the laws which control the fall of particles are the same under all climates, the rate of fall only being affected by temperature. The essential nature of such processes is therefore azonal, although the forms arising from them include a subsidiary zonal element.

Wind action provides an example. The laws of aerodynamics are the same whatever the climate. What is variable is the size and the velocity of the eddies. Nevertheless, the geomorphic action of the wind varies with climate. One factor is the resistance offered by any plant cover; within forests the wind acts indirectly by way of windfalls which it produces, and around glaciers the wind modifies the distribution of snow, and so on. The material transported also varies with climate; it may be sand, silt, snow or aggregates of silt cemented by salts as on salt flats. The ground surface itself may also be modified zonally by purely physical factors; in some cold, humid climates the wind may move gravel and pebbles across ice surfaces which would not move in other conditions. In extremely cold regions the wind gives rise to stratified accumulations of sand and snow, known as niveo-aeolian deposits. Thus, wind action is essentially azonal, its effect being modified only by temperature and any modification being minor. Zonal effects are present in the environments in which the wind blows. This is not the standardised laboratory environment but a geographical one whose characteristics are a function of climate and are therefore zonal. These zonal characteristics dictate the nature of the ground surface which may vary from hard, encrusted surfaces, to more or less indurated sand or silt layers which provide the wind with its load, to an ice crust or a plant cover with a variable surface roughness which either facilitates movement of particles by the wind or, on the contrary, limits its movement entirely.

Coastal processes provide a similar example. The dominant process, that of waves breaking on the shore, is certainly azonal. The zonal factors here are perhaps less pronounced even than in the case of the wind. But again it is the local environment which introduces zonal factors. Rock disintegration under subaerial attack and local fluvial sediments provide most of the corrasive material used in wave attack so that the nature of this material varies according to the morphoclimatic region. Some shorelines in the Arctic are buried beneath alluvial deposits made up of frost-shattered material transported by streams. Shorelines in the humid tropics, where biochemical weathering is dominant, rarely have any pebbles and are made up mostly of sands and muds. The beaches in the middle latitudes, where mechanical disintegration is more important, are frequently rich in pebbles although these may be relics of earlier processes. In some environments biotic activities may counteract mechanical processes. In tropical oceans, for example, constructive organisms are very numerous and extremely active, producing fringing and barrier reefs which serve not only to modify the coastline but also to affect the mechanical processes which go to fashion it. The coastline protected by a barrier reef is unaffected by swell. So far the comparison with wind action on the land surface contains no difficulties, for disintegration provides the material used by both wind and waves. The vegetation cover affects wind action in the same way that constructive organisms affect coastal morphology. In the higher latitudes, however, an additional point emerges; here, the wind blows outward from the land in all cases while wave action is absent or extremely rare because the shorelines are perennially frozen. In this case it is clearly not just a question of a minor modification in the efficacy of an azonal process so much as a drastic change in the geomorphic mechanisms, such as the Antarctic ice shelves which are true coastal nivation forms resulting from the accumulation of snow on top of sea ice. One difference from wind action, then, is that the glacial zone is not affected by wave action on the coast. This distinction is catered for in the term polyzonal. Wave action may be considered polyzonal as it affects only some of the world's morphoclimatic zones, a characteristic in no way related to variations in the nature of the process itself.

In the case of running water the zonal component is even more marked. To make this point the clearer, let us briefly review those points which running water holds in common with the two geo-morphic agents already discussed. The laws of hydrodynamics are universally valid, climate influencing only the size of the parameters; however, the geographical extent of the influence of running water depends on zonal factors because drainage characteristics are the result of the interaction both of direct and indirect (soil type, vegetation) climatic factors. The geographical extent of running water is even less than that of coastal wave action. Two climatic factors influence this distribution: low temperatures (as is also the case with wave action) and dry periods. In the coldest zone of the earth drainage is in solid form, with glaciers advancing and modelling the relief in a way quite different from that associated with running water. In the most arid regions, on the other hand, there may be no drainage at all, precipitation being inadequate to produce a drainage network. In some climatic zones running water plays a part in the development of relief with an azonal character, while elsewhere it is entirely absent. The term polyzonal (or plurizonal) is particularly appropriate to this important intermediate condition between zonal and azonal effects.

We may now turn to the relationships between zonal and extrazonal phenomena.

Extrazonal phenomena are similar to zonal phenomena except that they occur outside their own zone. They may be rather different from the corresponding zonal phenomena because they occur in a different general geographical environment. The subdivision of glaciers into local (or mountain) glaciers and ice sheets is a case in point, conditions, ice temperature, modes of alimentation and ablation being different in each case.

There are sound reasons for considering extrazonal phenomena and their formative factors. These factors fall into two major groups: extension phenomena and the environmental characteristics.

Extrazonal features due to extension or survival

Extrazonal phenomena may be due to extensions of certain geographical features either in space or time. The first group of extrazonal features results from spatial extension. A good example is the persistence of fluvial morphology along the Nile in Egypt. The waters cutting the Nile valley owe their origin not to precipitation falling in the deserts through which the river flows but in a much more humid climatic zone, namely the equatorial zone and the mountains of Ethiopia. The persistence of the Nile valley in the middle of the desert, then, is the result of a supply of water which is too great to be absorbed in the long desert crossing. Such conditions modify the fluvial processes themselves. With the loss due to irrigation, evaporation and the absence of any important local affluents, there is a gradual decrease downstream in the discharge of the river. This contrasts with the situation in a zonal river, if we can leave karstic situations aside.

Glaciers also provide examples of extrazonal phenomena due to extension. Mountain glaciers frequently extend far into forest areas. The outlet tongues of the inland ice sheet pass through rock desert and tundra on their way to the sea. This was the case with the Pleistocene ice sheets whose terminal zones extended many hundreds of kilometres beyond the climatic limits of the glacial zone. As with rivers, such extensions beyond the zone of origin are associated with changes in the morphogenic processes; the melting of glacier tongues or the edges of ice sheets has neither the same quality nor the same geomorphic result as in the glacier zone proper. Debris accumulates below the melting ice and running water is abundant, at least in the warm season, as it excavates channels and hollows, cutting up the ice itself and removing its load. The result is a morphology quite distinct from that of the strictly glacial zone.

Such extrazonal phenomena are normal to glaciers but as far as streams are concerned, they are typical only of the larger rivers. However, in both cases the effects are represented spatially.

There is yet another example of extrazonal phenomena which in this case represent temporal extensions and so may be termed survivals. These are certain old ferruginous or bauxitic duricrusts which formed under a different climate from that of the present yet continue to grow at the base if well drained. Given that this explanation is valid then the duricrust may be regarded as a typical survival and the present duricrust processes are extrazonal. Another good example, already referred to, is the present distribution of permafrost beneath the Siberian taiga and deciduous forests. This extrazonal permafrost yields a distinctive suite of periglacial forms termed cryokarst by Russian scientists. This process is still morphologically active.

By the study of morphologically active surfaces, therefore, it is possible to distinguish extrazonal extension phenomena of both temporal and spatial kinds on the one hand and survivals no longer undergoing active development on the other. The latter kind of survival persists in a *residual* way, therefore. These are *relict forms* derived from previous conditions. They indicate a time-lag in the adjustment of the morphology to subsequent change in climates (*climatic residual forms*) or the termination or rapid deceleration of tectonic processes (*tectonic residual forms*). For example, the volcanic relief of the Cantal and the periglacial forms of the Paris Basin are both relict forms. In contrast, an extrazonal form is a *present form* still evolving, even if more slowly than in the past, such as duricrusts or cryokarst.

Of course, spatial and temporal survivals are not mutually exclusive. Both are found, for example, in the Greenland ice sheets. Here temperature measurements demonstrate that the ice is colder at a depth of 125 m than it is at the surface, suggesting a time-lag in the adjustment of the ice sheet to the postglacial climatic amelioration. Accordingly, part of the Greenland ice sheet may be regarded as a relict form which, however, is capable of producing extrazonal outlet glaciers.

Extrazonal features arising from the geographical environment

Some landforms of extrazonal type may be too small to be the result of extension or survival, in which case they owe their origin to the geographical environment.

Some of the minor geomorphic forms, such as the badlands of semi-arid regions, are the direct result of the bedrock lithology on which they develop. Badlands are zonal in arid regions but extrazonal in more humid regions, whether these be temperate or tropical. In humid regions however, they develop only on ground stripped of vegetation and not under forest or even on continuous grassland cover. Moreover, the surface must be made up of argillaceous or marly material and not subject to mass movements. In these conditions, severe rain showers readily generate gullies and badlands as may be seen in the southern part of the French Alps and as far north as Burgundy (near Chatillon-sur-Seine). Although not very common, badlands also occur in the periglacial zone, where they again depend on the nature of the bedrock. Here they are favoured by sands rather than by clays or marls. In northern Canada badland forms contrast strongly with the slopes due to gelifluction which tends to destroy the rills arising from rainwash. Another example of an extrazonal process is the solifluction of clays and marls in a temperate climate. For example beds of limestone, sandstone or basalt overlaying waterlogged clays or marls may cause downslope movement of the latter. Such deep solifluction—an unusual use of this term which is not to be found in the English language literature of geomorphology, where it specifically denotes mass movement of heterogeneous debris downslope under the action of freeze–thaw and high pore-water pressures—without frost beneath a developed vegetation cover differs in some ways from gelifluction on permafrost although its general characteristics are the same. The extrazonal nature of the process is made clear by the terminology adopted here which restricts the term *gelifluction* to soil movement on permafrost and the term *solifluction* to all other varieties of slow soil flow whether in periglacial regions or in the tropics.

Extrazonal features may also arise from particular topographic conditions. This is true of some groundwater duricrusts found in the continually humid rather than in the seasonally dry tropics. In the rain-forests, these extrazonal duricrusts are found on valley floors where the fluctuating water table lies not very far below the surface.

Tectonic factors may also cause extrazonal extensions of certain landforms. For example the *sebkahs* (salt flats) around Oran in western Algeria are found in a region normally too moist to permit their development. Under the present climate an intermittent exoreic drainage would be expected rather than an endoreic one. There is some evidence that the salt flats are unstable but they persist due to the continuation of tectonic activity which produces synclinal depressions which, in turn, prevent the rivers from reaching the sea. Thus in this case, the *sebkahs* are extrazonal and the result of present-day tectonics. This may have been true of some salt deposits in the geological past.

Finally, altitude may cause extrazonal extension. Above a certain height periglacial processes and glaciers may exist in temperate and tropical mountains. These have sometimes been regarded as extrazonal but, in fact, the environmental conditions are quite distinct and they require separate consideration.

LATITUDINAL AND ALTITUDINAL ZONATION

Latitudinal and altitudinal zonation have often been confused, the successive altitudinal zones on mountains being regarded in the same light as the succession of zones with movement from the equator to the pole. By this reasoning, rise in elevation in the tropics should produce a zonation which is tropical, temperate, periglacial and then glacial. This is at best a very rough approximation of reality. There are important differences between the two which led Flahault, the geobotanist, to use the term *étage* as early as 1900 in preference to *zone* for mountain areas. These two concepts should not be confused in geomorphology as they are in some other natural sciences.

DISTINCT CHARACTERISTICS OF VERTICAL ZONATION

Vertical zonation is the result of the change in temperature and precipitation with height. The higher the altitude, the thinner the air: accordingly, its heat capacity falls and its transparency generally increases. The daily air temperature decreases while that of the ground surface increases. As the air is thin so the solar radiation is more intense. One result of this is that rocks are subjected to much greater thermal effects in mountains than they are in lowlands, effects which become more marked with increased altitude. In middle latitudes, for example, rock surfaces may reach a temperature of around 20°C in the early part of the afternoon and −5°C at night between altitudes of 2500 and 3000 m. Even greater ranges are to be found at the higher elevations in the Mediterranean and in the tropical zone. The amount of cloud cover is, of course, one influence on such temperature ranges, so that in cloudy climates, by way of contrast, it is the fall in the air temperature range that becomes important because rocks then remain at temperatures close to the air temperatures. This becomes especially important on less elevated mountains where the air is not significantly rarified, as in the Vosges, the Harz Mountains and the Scottish Highlands and on humid maritime islands. The variation of precipitation with altitude is a rather different phenomenon. Starting at sea-level, precipitation progressively increases with altitude but then reaches a maximum above which it tends to decrease. On mountains that are sufficiently high, therefore, a zone of maximum precipitation and cloudiness is to be found. As the air rises it cools by expansion until the dewpoint is reached and condensation occurs. The air rising from adjacent plains reaches the dewpoint at progressively lower altitudes on the mountain sides as the water vapour content of the air increases. Accordingly, every climate has an optimum level of condensation corresponding to the mean elevation at which rising air condenses. This level of maximum precipitation is to be found in all climates even the most arid. It is at the source of major differences between mountain morphogenic processes and those of the lowlands. Even in the arid regions there is to be found at least one altitudinal zone in the mountains which possesses an integrated drainage system, if not perennial streams. Good examples are provided by the Andes in northern Chile, the Ethiopian Highlands and even the Ahaggar in the Sahara.

The *topographical* element in vertical geomorphological zonation is naturally very important. The fundamental difference between latitudinal and altitudinal zonation is that the former affects vast regions and only very few of the world's rivers have basins large enough to enable them to flow through several morphoclimatic zones. In mountain zonation, on the other hand, the existence of several distinct morphoclimatic *étages* in the same moderately sized basin is the rule. A single slope may often stretch across several *étages*, so that the footslopes are covered with forest while summits carry Alpine meadows or even perpetual snow. While lowland rivers differ from one morphoclimatic zone to another, mountain streams, which are always to some degree hybrid, resemble one another through all the *étages*. A major reason for this is that general steepness of slope favours surface runoff: the discharge–drainage basin area ratio or *module* and the proportion of the total precipitation evacuated in the streams, or the *drainage coefficient*, in mountain streams always have higher values than those in the streams of the adjacent plains. Accordingly, mountain streams are that much more dynamic and their activity is expressed in a greater drainage density and a more violent flow. It follows that the geomorphic activities of such streams is considerable, causing the term *torrent* to be applied to them (the Latin *torrens* signifies a stream course that dries up during the summer heat). Variation in discharge is further increased by the generally steep slopes, so that mountain areas are characterised by active dissection and rapid slope evolution. Generally, then, the mechanical processes are more active than they are in the adjacent lowlands.

The three basic factors of temperature, precipitation, and topography serve to distinguish vertical and latitudinal zonation and bestow on morphogenic systems a suite of characteristics which marks them off from those of the neighbouring lowlands. However, this is a rather one-sided view, for latitudinal zonation also influences vertical zonation.

THE EFFECT OF LATITUDINAL ZONATION
ON VERTICAL ZONATION

Regardless of altitude, some features characteristic of the zonal climate are found in all the world's mountains, and might well form what could be called the 'basic climate'. They ensure the persistence of some zonal originality at the several mountain *étages* and are most simply expressed in the temperature régime. For example the dominance of diurnal temperature oscillations over seasonal ones in the equatorial regions, so typical of the lowlands, occurs at all altitudes and is expressed in a special glacial régime.

On this general question two considerations must be mentioned. Firstly, mountain zonation is not identical in all latitudes and, secondly, there are differences between vertical zones and their equivalent latitudinal zones.

The effect of latitude on vertical zonation

The principal consideration here is the height of the maximum precipitation belt above sea-level. To this may be added the effects of site, for, within the same mountain range, precipitation maxima occur at higher altitudes in the interior than in the outer, marginal mountain areas, for example Pre-Alps, 1300 m, Mont Blanc, 2500 m. Generally speaking, precipitation maxima occur at lower levels in humid climates. For example, in Hawaii which has no clear dry season, the condensation level is only 700 m, while it is 800 m in Java, 1300 m on the southern slopes of the Himalayas and the Black Forest and 1400 m in the Aigoual (Cévennes) and the Sierra Nevada of California. Thus, the precipitation maxima occur at lower altitudes in the humid tropics than they are in the temperate zones, while they appear to be at their highest in the tropical deserts.

The vertical interval between the 0°C mean isotherm and the precipitation maximum increases from the poles to the equator. They almost meet in mountains at the Arctic circle as well as in some of the higher massifs of the Alps, as is the case in the Mont Blanc massif. This situation provides optimal conditions for glacier development, for part of the mountains lying above the snowline receive enough precipitation in the form of snow for the low temperatures to preserve a great deal of it so that it persists through the warm season. By way of contrast, there is a considerable altitudinal interval between the 0°C isotherm and the maximum precipitation belt on the equatorial mountains. This reaches more than 4000 m in the Peruvian and Bolivian Andes and in East Africa. As the upper levels do not receive a great deal of precipitation, glaciers do not develop despite the generally low temperatures. Moreover, inadequately nourished glaciers do not extend far below the snowline. Thus there is a unique desert or semi-desert belt between the upper forest limits and the snowline which results from the generally low precipitation. The absence of a marked winter season resulting from the latitude means that snowfields vary little in their extent throughout the year. Outside the snow and icefields, then, the regolith is left unprotected against large diurnal temperature variations, so that this high altitude tropical desert is susceptible to the periglacial processes of needle-ice (pipkrakes), freeze–thaw and solifluction. Major differences between this environment and periglacial *étages* of mid-latitude mountains include the following:

(1) adequate moisture in the latter which allows meadow vegetation to develop which, in turn, impedes the action of solifluction, leading to lobe and terrace forms;

(2) a thicker snow cover which serves to reduce the direct attack of frost during the winter;

(3) generally higher precipitation and more limited soil temperature ranges producing a quite different process balance; and

(4) the glaciers are much better supplied with snow and descend much greater distances below the snowline than they do in the tropics so that a greater proportion of their tongues is climatically extrazonal.

Latitudinal effects on vertical zonation are not, however, restricted to the tropical and temperate zones. Friedland (1951), in a systematic study of vertical zonation in Russian soils, recognised a boreal type which is well developed in the mountains bordering the Sea of Japan. This zonation is made up of forest and podsolic soils with tundra above, and it would seem that these montane tundras are the equivalent of the Alpine meadows of mountains in temperate climates. Such meadows are absent from this boreal zonation in the same way that tundras are almost entirely absent in temperate mountains.

Contrasts between vertical zones and their equivalent latitudinal zones

Some comment is required on the general similarity between mountain glaciers and the periglacial zone below them and the glacial and periglacial zones of high latitudes. Analogies between these two have frequently gone too far in the past. For example, conclusions based on the study of one or two glaciers in one of the environments has been too readily applied to the other. This serves only to distort the facts. Glaciological work, notably that of Ahlmann, has distinguished temperate from polar glaciers with their two quite different régimes. Moreover their morphogenetic effects are quite distinct as the effects of mountain glaciers may be quite different from the effects of ice sheets.

Comparable differences exist between the latitudinal and the altitudinal periglacial zones. They have been stressed, perhaps somewhat enthusiastically, by Troll. In the polar periglacial zone all the relief is zonal. Even the stream courses have their general erosional effect, especially on the development of slopes. In mountains, however, stream courses are in no way distinctive and retain their torrential azonal characteristics. Here, only the slopes develop under the control of periglacial processes and then only partly so because the laws which govern these processes do not affect the general denudation. In the mountains a slope's base-level is usually reduced by the incision of torrential streams while in the Arctic it may actually be raised under apparently similar conditions due to gradual infilling of the valley floor. Such deposition and base-level rise is much rarer in mountains. A further factor affecting landform evolution in periglacial mountains is *exposure*, and the influence of this factor may raise or lower the zonal limits of two adjacent slopes by up to several hundred metres. As is well known, in a temperate climate the shady side or *ubac* helps to preserve snow patches almost throughout the summer and is subject to freeze–thaw action, while the sunny side or *adret* at the same altitude may be covered by Alpine meadow vegetation. Great contrasts in precipitation may also arise from the particular orientation of the mountain massif: the humid north-west-facing slopes of the Middle Atlas of Morocco have a grass cover, while those facing south-east, overlooking the Moulouya valley, are surfaced by frost rubble and scree. The importance of exposure varies from one latitudinal zone to the next although some have said that it is limited to middle-latitude mountains. This is not true, for Tricart has shown it to be important in the equatorial Andes of Venezuela, Dollfus has done the same for the Peruvian Andes and Raynal for the subtropical semi-arid Atlas Mountains of North Africa. In the Venezuelan Andes, an important factor is the development of a cloud cover due to orographic factors during the daytime, so that south-east and eastward-facing slopes receive insolation before the clouds develop so that any overnight snow melts while the slopes that face north-west and west receive little sunshine. These slopes are very moist and covered by moss forests and, at higher altitudes, glaciers which are fed in part by condensation of of atmospheric moisture on the firm surfaces.

The more consistent atmospheric circulation of the tropics emphasises contrasts in moisture levels on opposed flanks of mountain ranges. Here windward slopes are humid while leeward slopes are dry. In Venezuela, windward slopes are overgrown by forests draped in mosses and algae while the leeward slopes have xerophytic shrubs and bushes. As might be expected, the leeward slopes suffer intense runoff erosion, while the main denudation process on the windward side is creep.

In Morocco, differences in exposure result both in contrasts in temperature (as in middle-latitudes), and contrasts in humidity (as in low latitudes). Such contrasts naturally attenuate as the cloudy climates of higher latitudes are approached, for differences arising from direct insolation become minimal. Friedland (1951) has stressed the effect of the wind in regions around the Arctic Circle. In the winter, wind may sweep the ground clear of snow so that cold waves may penetrate deeply into the ground. Here, it is the valley floors which are covered by tundra and the valley sides by taiga. The least vegetation cover is found on the hillcrests due to their greater exposure to the wind. Aspect plays some part here though it is less marked than in the middle-latitudes: peat bogs commonly develop on the shady side of ridges.

One zonal characteristic of latitudinal type found at all *étages* of tropical mountains is the virtual absence of distinct cold and warm seasons. The morphological processes, then, have a diurnal rather than a seasonal rhythm.

It follows from all this that the differences noted between montane and polar periglacial morphogenetic systems cannot be explained in simple extrazonal terms. Mountain environments appear to be a special case. True, they share many characteristics with the lowland periglacial zones but they do have a great variety and many quite specific features. The nature of the processes themselves and the boundaries of the vertical *étages* vary from one slope to the next as a result of the disposition and form of the mountains. It may be concluded that the lack of systematic work such as that attempted by Friedland, on the *étagement* characteristics in the world's mountains is a great gap in our knowledge.

Used with discrimination, the zonal concept facilitates the analysis of the world distribution of the morphogenic processes. Moreover, given that allowances are made in each case for the particular as against the general case, it makes possible the classification of the world's morphogenic systems and the delimitation of the great morphoclimatic zones.

MORPHOCLIMATIC ZONES OF THE EARTH

Delimitation of the world's morphoclimatic zones should be based primarily on present-day phenomena. At the same time, paleoclimatic factors should not be neglected, for relict landforms can be numerous and failing to take them into account raises the risk of misunderstanding the morphology and its mode of origin. Let us now consider the criteria to be used in such a subdivision of the earth. Such an exercise is fraught with difficulties for which there are three main reasons.

(1) Systematic studies on this subject are inadequate in number. Some efforts have been made to classify the continents according to morphoclimatic types but these have rarely been accompanied by accurate maps. For the most part, these types and their geographical limits should be mapped in the field but, on the whole,

this has not been done. It is for this reason that the extent of the morphogenic systems of central Asia, South America and Africa are so poorly known. Work on a broad scale (see Tricart, 1958) aiming to map present morphogenetic systems as well as the results of paleoclimatic changes, is still exceptional. Yet it is this very approach which is demanded.

(2) Morphoclimatic classification cannot be based solely on climatic data because climate affects the landforms in both a direct and an indirect way. This is true even if the climatic parameters are used in an ecological way as are those of Thornthwaite. Moreover, while the major vegetation formations may share common boundaries with some morphoclimatic units, these do not coincide, so that a map of vegetation formations will not serve. There are not 32 different morphogenic systems as there are 32 different types of climate under Thornthwaite's scheme. Adaptation of a simple biogeographical boundary, such as that between the pine forests and hardwood forests of the south-eastern United States, for morphoclimatic purposes is not possible. By the same token, it is not acceptable to group the 'parkland savanna' of the Sahel with the short grass savanna of the Magreb or with the steppes of central Asia; neither can the coniferous forests of Scandinavia be put in the same category as the boreal conifer forest of central Siberia.

(3) Another source of difficulty in morphoclimatic subdivision is to be found in transition zones and the more subtle changes. Morphoclimatic boundaries are clearly marked only where there are strong orographic contrasts. Very often, however, transition zones cover some hundreds, or even some thousands of kilometres. This is true of Central Europe, for example, with the change from the maritime climate with mild winters to a continental climate with extremely cold winters. It is true also in the Arctic with the transition from rock desert and tundra to tundra and forest. Enclaves and exclaves (analogous to the ecotones of biogeography) occur with variations in exposure, lithology and human and animal interference.

Accordingly, it is impossible at present to construct anything more than a provisional map of the earth's morphoclimatic zones. The most ambitious attempt at such a classification is that of Büdel (1948). Unfortunately, this classification contains inconsistencies; for example it treats submarine relief (which is azonal) on the same basis as subglacial relief and the various subaerial morphoclimatic zones.

Except in the case of the high altitude (cool) deserts, no account is taken of the extrazonal character of the world's highlands. Also, it contains morphoclimatic subdivisions which are difficult to sustain. For example, the regions of permafrost extend into three categories (the frost debris zone, the tundra zone and the nontropical zone of mature soils with permafrost), while the Siberian boreal forest regions with permafrost are put in the same category as the maritime temperate forests and the steppes. The Mediterranean regions, on the other hand, are regarded as a separate, distinctive zone. There appear to be fewer differences, however, between degraded Mediterranean areas and the steppes, or between Mediterranean regions which have retained their forest and the middle-latitude forests, than exist between forested areas of the oceanic temperate zone and the zone of mature soils with tjäle. Finally, the criteria underlying the sheetwash zone are inconsistent because this is, together with the middle-latitude and equatorial zones, a zone of mature soil development.

Accordingly, we find Büdel's classification unacceptable and, in its place, we propose a classification resting on two types of criteria, as follows.

(1) Principal divisions are based largely on the major world climatic and biogeographical zones.

(2) Each of these major zones is subdivided morphoclimatically on the basis of further climatic and biogeographical criteria, as well as paleoclimatic factors.

In areas of low altitude where there is no appreciable vertical zonation, therefore, the following zones may be distinguished.

The cold zone

This is where frost action is of the first importance. On the basis of the latter, the zone may be subdivided into (a) a glacial zone where runoff is mainly in the solid form of glacier ice, and (b) a periglacial zone with seasonal runoff in the form of water but in which ground ice is the predominant agent in the development of the interfluvial areas.

The mid-latitude forest zone

This is variously affected by man's activities, in which relict forms, notably those of Pleistocene glacial and periglacial origin, are very important. The differential effect of winter freezing (which is a major geomorphic process here) together with the paleoclimatic factor, produces the following morphoclimatic subdivision: (a) maritime zone with mild winters, limited frost action, and significant glacial and periglacial relict forms of Pleistocene age; (b) a continental zone with severe winters, strong frost action at present and during the Pleistocene, and commonly characterised by the persistence of Pleistocene permafrost; (c) a Mediterranean zone with dry summers in which relict periglacial landforms of Pleistocene age play only a minor part.

The dry zone

This comprises the arid and semi-arid regions of middle and lower latitudes, and has discontinuous vegetation of steppe, xerophytic bush and desert types and intermittent surface runoff. It is subdivided according to: (a) the amount of the water deficit underlying the three-fold division into steppe, xerophytic bush and desert; and (b) the winter temperatures which have an important influence on certain processes such as frost-shattering and the duration of any snow cover. This temperature factor is the basis of the distinction between middle-latitude and tropical–subtropical dry regions.

The humid tropical zone

This is characterised by persistently high moisture and temperature levels and perennial stream flows. It may be subdivided into the savannas and the forests on the basis of the seasonality of the rainfall, the annual rainfall amount and the related factor of the density of the plant cover.

(1) The savannas of the seasonally wet tropics have a moderate degree of plant cover which is particularly susceptible to fires kindled by man. They are subject to widespread overland flow and severe chemical weathering during the wet season. In regions which have not been subjected to fires for many centuries, the savannas are represented by the *campos cerrados* and the deciduous tropical forests.

(2) The tropical rain forests receive abundant rain throughout the year and have a luxuriant vegetation cover which is not susceptible to burning. It is here that both chemical and biochemical weathering reaches its greatest intensity. A subsidiary subdivision might be

made between the evergreen forests and the semi-deciduous forests which display some seasonal loss of leaves, as this influences runoff.

Given our present knowledge any geomorphological region on earth can be accommodated in one or other of these zones. We now turn to the characteristics and extent of these zones and their sub-divisions.

In the *cold zone* the incidence of frost is the critical climatic factor. It may act directly (that is, by inhibiting vegetation or by the pro-duction of contraction cracks in the regolith) or indirectly by changing water into ice. Frost action markedly affects the action of the azonal processes—wave action, wind action, and running water. Accordingly subdivision of the cold zone is based on the severity and incidence of frost.

The glacial zone is found in most parts of the earth where a thermal deficit is such that snow persists from one year to the next, so that glaciers grow. The subglacial relief produced in this zone is quite distinctive and differs enormously from that associated with subaerial processes. As we have seen, while mountain glaciers and ice sheets have certain landforms in common (moraines, *roches moutonnées* and striations), they possess important geographical, thermal and dynamic differences. In middle-latitude mountain glaciers, for example, torrential summer meltwaters make possible the evacuation of englacial debris from the glacier snouts. Such glacifluvial deposits accumulate during retreat phases while end moraines develop mainly during glacial advance. On mountain glaciers near the equator, on the other hand, glacifluvial deposits are extremely limited but the end moraines are enormous. Push moraines, made up of frozen debris, are not found at the snouts of mountain glaciers in middle-latitudes, but they are common on the edges of ice sheets. Coinciding with the glacial margins, the borders of the glacial zone are quite evident. As we have seen, they are not strictly climatic for ice has its own dynamics. The margins of the glacial zone do not accord precisely with present climatic limits because some ice still survives from the Pleistocene, to judge from the ice temperatures in Greenland quoted above.

The periglacial zone is characterised by recurrent freezing, with freeze–thaw cycles which are generally more rapid than in the glacial zones. Here, snow does not persist all the year round. The thaw produces meltwaters which feed the streams which play a part in fashioning the relief. The two outstanding geomorphic processes in this zone are frost-shattering of rock and geliturbation of debris. The latter is especially important on moist slopes rich in clay.

The periglacial zone contains several subdivisions which are based on three criteria: the frequency of freezing both under the present climate and under climates of the past; the insulation against freezing provided by the vegetation cover; and the annual precipitation.

Carl Troll distinguishes diurnal and seasonal freeze–thaw alternations. The diurnal type is known from some maritime areas such as Iceland and also from tropical highlands, so that it is areally of minor importance. On the other hand, the seasonal freeze–thaw type is quite widespread and includes parts of the mountains of middle-latitudes as well as Greenland, Alaska, Canada and Siberia. The important thing, however, is the recognition of differences in the duration of *ground freezing* (daily, seasonal or perennial) as this is a major determinant of the morphogenic processes. While a few days' freezing may affect only a thin layer of regolith, seasonal frost penetrates several decimetres or even one or two metres. Permafrost, of course, radically modifies the runoff of an area and so has a marked morphological effect. As we have seen, the thickness of the snow cover may be important in that it may act as a protective insulating layer against frost action.

The protection offered by vegetation clearly varies according to the vegetation community which, in turn, is related to the climate. In this instance, Büdel's distinction between rock desert and tundra is useful. In the tundra there exist soils, albeit thin and skeletal, and also bogs both of which respond to frost action quite differently than does raw regolith, so that highly distinctive forms may occur, for example string bogs. Generally speaking however, tundra acts to impede geliturbation and gelifluction as may be seen in its distinctive microrelief of lobes, hummocks and *palsen*. This is also true of mountain meadow vegetation. Finally, of course, vegetation limits wind action which has free play in the rock deserts.

The boundary between the cold zone and the *middle-latitude forest zone* cannot be drawn with any precision. While there is no doubt that all tundra areas are to be included in the cold zone, the boundary between the tundra and the boreal forest zone is irregular and frequently uncertain. This transition is identical with Rousseau's peri-arctic zone which is so clear on the ground in Newfoundland and northern Quebec. Frost action is the dominant morphological process in the open forests bordering the tundra, so that this zonal boundary is clearly no more than a broad transition zone. Forests growing on discontinuous patches of permafrost are sufficiently large to be mapped separately.

Between the tundra and the dry zone is an important series of forest formations. These middle-latitude forests make up an extensive belt in Eurasia stretching from the Atlantic to Lake Baikal only to re-appear in the Amur Basin, Korea and Japan. In North America the forest extends over the eastern half of the continent from Texas to Labrador and from Florida to the Yukon valley where it seems to give way to coniferous forest on permafrost. On the west coast of North America the forests extend from northern California to the Alaska peninsula. In the southern hemisphere they cover only limited areas: the Pacific coast of South America south of Santiago de Chile, the coast of Natal, the eastern Australian coast all the way to the Tropic of Capricorn, Tasmania and New Zealand. Forests that are transitional between tropical and middle-latitude types occur especially on the eastern sides of the continents, where the climate is moist enough to interrupt the dry zone. This is true of southern China, southern Japan, south-eastern United States and eastern Australia. These *subtropical forests* occur in areas where Pleistocene periglacial action was felt only on the higher mountains. In fact, they are transitional between the humid tropical forest regions and the deciduous mid-latitude forests of, for example, the north-eastern United States, northern Japan and the far eastern USSR. The warm humid summers in this zone give rise to a deep-weathering profile which is quite similar to that found in the humid tropics, as Lautensachs (1950) and Krebs have demonstrated in Korea, although Korea is perhaps less typical than is southern Japan.

In the *mid-latitude forest zone* the characteristics of morpho-climatic significance are a fairly dense vegetation cover which provides protection from the mechanical processes and only moderate average temperatures which modify the rate of chemical reactions. In an undisturbed condition, this is a zone of rather limited morphological activity in which landform evolution is distinctly slower than in the two adjacent zones, the cold and the dry. Very steep slopes can be maintained under a forest cover as in the Apuane Alps north of Pisa, in Japan and in the fjordlands of British Columbia or southern Chile. As decomposition of humic material is slow a thick forest litter covers the ground. This, together with generally good soil structure, greatly reduces overland flow which may be quite negligible. Indeed, creep is frequently more notable than overland flow.

The generally weak morphological processes, of physical, mechanical, chemical and biochemical types, favour the preservation of relict Pleistocene landforms. A great deal of the morphology of this zone, is, in fact, relict. The importance of mechanical processes is dependent on the severity of slopes and their lithology. Steep slopes form by rock-fall and landslides, by the development of solifluction lobes and by incision of first-order streams. The gentler slopes have deep soils which are a testimony of the minor importance of sheet erosion. The main types of mass movement appear to be those due to slow downward movement of debris as a result of compaction arising from the loss of material in solution, the activity of burrowing animals and the effects of the wind in uprooting trees. Biochemical processes do not act as quickly as they do in the tropics; the soil minerals show less advanced weathering and the regolith is not as thick, averaging about one metre. The loss of dissolved matter is less also. In areas of low relief, the total of solid load transported appears to be extremely small under undisturbed conditions.

However, the climates within the mid-latitude forest zone are sufficiently varied to affect the balance between individual morphogenic processes such that three subdivisions may be recognised.

An oceanic zone which has a limited annual temperature range but a large humidity range. This is well developed in western Europe from Norway to the Pyrénées and extends into the interior of the continent as far as Poland. It is also found in British Columbia, southern Chile, Tasmania and New Zealand. Frosts are moderate and do not last very long. Frost rarely penetrates to bedrock. Desiccation of the ground is rare as summers are generally rainy, so that variations in stream discharge are limited and mechanical processes, mainly the result of living organisms, are also limited. Chemical erosion is dominant and is active throughout the year, inhibited only by low temperatures. Humus on the ground surface tends to become acidic thus aiding chemical attack of rocks such as granite.

A continental zone, characterised by a much less equable climate— heavy showers, even when annual precipitation totals are small, together with severe winter cold. Mechanical processes are more important here than in the oceanic zone. Frost penetration is deeper, often reaching bedrock and affecting runoff in a large-scale way for several months of the year. Peak runoff associated with seasonal thaw and brief periods of thaw during winter is typical as far as western Siberia. The morphological results of this are striking, especially as they are increased by snow meltwater. Sheet erosion and gullying are common so that the dissection differs from that in the oceanic areas. Thunderstorms also help to maintain the landforms produced by runoff during the thaw, although they are rather less important morphologically. Chemical erosion is limited by winter frost and by large volumes of water which escape by surface runoff rather than by infiltration. Accordingly, chemical action is less notable than in the oceanic areas and mechanical processes are correspondingly more important, especially as slopes become steeper. Nevertheless, the mechanical processes are inhibited by forests which cover quite steep slopes, certainly on the consolidated rocks. Factors such as these help to explain the considerable differences between the erosional landforms of northern Japan or Korea and those found in north-western Europe. Slopes may be very steep (averaging between 20 degrees and 35 degrees) and more rectilinear. Other differences exist but these are the result of past climates and are discussed below.

A warm temperate or subtropical zone typical of areas with a Mediterranean-type climate. At low altitudes, frost is rather rare and slight and only severe modification of the vegetation cover by man can render it a notable morphological process; otherwise it is limited to bare rock outcrops. The major morphogenic role is played by alternating wet and dry periods which result in volume changes in argillaceous soils, one result of which is an increased likelihood of landslides. Also, the seasonal precipitation régime makes the streams torrential and capable of moving coarse debris. Rain showers are commonly violent so that runoff tends to be in streams within a dense drainage network so that rapid dissection results. Of course, erosion is a function of the annual precipitation. In very wet regions, such as the Apuane Alps or the Ligurian Apennines, slopes which have been kept steep (between 30 and 40 degrees) by a vegetation cover have been severely scored by a secondary stream system of very high density. The result is comparable in both areas although one is underlain by schists and the other by limestones and marbles. Such intense runoff also occurs in climates lacking a well-defined dry season, for example the southern Appalachians, southern Japan and the east coast of Australia. Runoff is facilitated in this subtropical zone by the more rapid decomposition of humus arising from higher average temperatures. Moreover, soils are generally thinner than in the oceanic and continental zones (especially the A-horizons) so that their protective role is less. Being adapted to a dry season, the forest cover in Mediterranean regions is not as dense as that found in cool temperate areas. Trees are not as tall and leaves are thick and not so numerous, often being replaced by needles. These characteristics, especially when reinforced by man's degradation of the forest, result in a reduction in the protective capacity of the vegetation and a general increase in splash erosion.

Current geomorphic processes, however, explain only a part of the morphology found in the mid-latitude forest zone. Two other factors are of considerable importance.

Paleoclimatic influence

The first of these is the paleoclimatic influence which is expressed by the presence of landforms which differ greatly from those actively developing at present. The mid-latitude forest zone in Europe and North America was invaded by large Pleistocene ice-sheets so that glacial erosional surfaces, moraines and preglacial deposits are widespread. Beyond the former ice margins an enormous area suffered a periglacial morphogenic system and gentle slopes (between five and 15 degrees) were generated by the process of solifluction. Present-day subtropical regions, such as the Mediterranean, the south-eastern United States, southern Japan, Natal, the North Island of New Zealand and the east coast of Australia, were rather differently affected by this Pleistocene climate and this difference is of assistance in the differentiation of these regions from the maritime and continental zones.

Anthropogenic processes

The second factor acting upon the present landforms is the anthropogenic or those morphogenic processes induced by man. This has been quite severe in the northern hemisphere, especially in the Mediterranean basin and in the continental zone because the mechanical processes have greater potential there. The degradation of the Mediterranean forests into the *maquis* and *garrigue* as a result of overgrazing has completely altered the morphogenic system on the non-cultivated land. Similar results have been produced in the space of a few decades in California and Western Australia. In the continental zone man's activities have resulted in the extension of gullying and the retreat of the forest in the cultivated steppe, especially in European Russia, the Ukraine and northern China. Similar erosion and loss of sediments has occurred in less than a century in the United States.

The combination of paleoclimatic landforms, current morpho-climatic processes and anthropic effects goes a long way to explain the range of landforms to be found in the mid-latitude forest zone. It is not surprising, given this complexity, that landform components such as man-induced gullying, have been singled out from this complex, and used to build an artificial system of so-called 'normal' erosion.

With distance from the mid-latitude forests towards the continental interiors, increasing aridity results in the gradual thinning out of forest and its replacement by less dense plant formations which set up a less efficient barrier between the atmosphere and the lithosphere. The forest merges into sclerophyllous bush (*maquis, garrigue* and chaparral) or wooded steppe, then grass steppe and desert. The precipitation becomes less in total and more irregular; annual totals themselves may vary from year to year by between 10 and 100 times. Much of the rainfall occurs as heavy showers. For example, at Cape Juby in the Spanish Sahara 57·3 mm of rain fell in three consecutive days. At Port Étienne in Mauritania, 2 mm of rain was received in 1912, and 301 mm in 1913, 300 mm of the latter falling in a single day. Clearly, low average annual rainfalls under such conditions do not result in negligible runoff erosion. On the contrary, when it does occur runoff can be brutally effective, especially as the vegetation cover is sparse and the soil is thin. Under such conditions infiltration is minimal and most of the water is shed by surface runoff. The streams cut the hillsides with gullies but disappear very soon after the showers end. This results in the formation of *bajadas* and pediments crowned by dissected inselbergs.

In such arid conditions, there is widespread interplay between water and wind action. Between periods of flood, the alluvial deposits suffer severe deflation, the coarser fraction being formed into dunes while the fines are removed in great dust clouds and may travel beyond the borders of the region. Saharan dust has been recognised as far north as Paris, for example. The relative importance of runoff waters and wind on the relief naturally varies according to the aridity and the nature of the vegetation cover. Three distinct environments may be recognised.

The subhumid steppe or short-grass savanna extends for vast distances north and south of the Sahara, in East Africa, around the Kalahari, in Asia Minor, Central Asia, Australia, the high plains of the United States and the Canadian Prairies, the Mexican plateau and the Argentine Pampa. In the undisturbed condition, the grass cover is effective in inhibiting mechanical erosion processes. Wind action is restricted mainly to stream beds and alluvial surfaces in the lee of which groups of dunes may develop. The main effect of the wind appears to be the accumulation of loess from both local and distant source areas. Thus, the great loess deposits of China were probably derived from the deserts of Central Asia. Runoff is commonly in the form of sheetwash, although it may become concentrated and set up a drainage network with incised gullies. As the natural vegetation cover is conducive to the formation of a good thick soil, the eroded materials are often well-structured loams. Aridity inhibits leaching and, even in hot climates such as the Sahelian zone of Africa, soils are neither ferrallitic nor particularly ferric. They tend to be brown and chestnut-brown in colour and even podsolic types may be formed. Where a long severe winter delays the decomposition of humic material, very thick A-horizons may develop as in the *chernozems* or black earths of the Russian steppes or the American Midwest.

As a result of the rich grazing potential of the steppe lands, they have been overgrazed and the plant cover, including the shrubs and trees, has been degraded. Thus disturbed, the soil lacks the necessary defence against erosion and is readily susceptible to mechanical processes. Linear flow, sheetwash and deflation are all serious in their effects and affect large areas.

The semi-arid regions display a discontinuous grass and shrub vegetation. Here again, rainfall is periodic and showery in type, so that runoff may be very localised. In contrast to the sub-humid steppe, integrated drainage systems are lacking or, if they exist, they are not active, being relics from past periods, such as the pluvial periods of the Pleistocene, when the climate was cooler and runoff was greater. For example, the Draa in Morocco is a continuous river course which runs from the High Atlas all the way to the sea, but the river-bed itself is rarely used by the same flood over the whole of its length. This type of climate gives rise to the fullest development of pediments and inselbergs. Rainfall is sufficiently great (averaging between 240 and 300 mm per year) for showers not to be unusual events, although integrated drainage systems cannot be produced. With a sparse plant cover and a thin regolith, overland flow is dominant. Running water of this type, then, is the main morphological process, being concentrated in the uplands and spreading out in sheets and rills at their foot to produce a highly individual landscape. Wind action is only subsidiary, reworking debris which has been produced by mechanical weathering and later distributed by sheetwash.

Some regions which have been semi-arid for a very long time have a distinctive vegetation of small trees and shrubs which are adapted to drought. Examples include the scrublands of Australia, the *caatinga* of north-eastern Brazil, the xerophytic scrub of southern Mexico, northern Venezuela, Colombia and north-western Peru. Between the strongly rooted plants the soil is quite bare. Sheet and rillwash are widespread and the development of a humic upper layer in the soils is prevented. At the same time, concentrated runoff is inhibited by the strongly rooted vegetation so that pediplains and inselbergs result, as occurs over much of Australia. Unlike the vegetation of the disturbed sub-humid steppe, that in semi-arid regions does not favour wind erosion.

Semi-arid regions, then, represent a belt of intense morphological activity around the edges of the arid regions and, due to the man-induced processes, they have been extended at the expense of the steppe.

The arid regions strictly speaking, are those which have no runoff. Such regions are, however, very few and those that do exist owe this character to particular bedrock conditions. They are always either sandy or stony, as are the Saharan *ergs*, high permeability preventing runoff. Where solid rock crops out some overland flow always occurs even in the driest of regions. In the Central Sahara and in the Libyan desert, some places such as In-Salah do not receive precipitation every year but, when it does fall, the wadis are filled. Dense networks of these wadis, carrying periodic flood-waters, occur widely in the Sahara other than in the sandy deserts. In the most arid areas, however, morphologically effective rainfall is rare and may not occur at all in some years. When falls do occur, their action is similar to that in the semi-arid areas. Accordingly, landform evolution is extremely slow. Moreover the visually more spectacular wind action also produces little result, for once it has removed the fines from the surface deposits it loses its power and its activity in the *ergs* is restricted to a constant refashioning of minor landforms. As a result, the location of the large *ergs* has remained the same for thousands of years and it may be that the detailed forms themselves are more stable than is generally believed.

The foggy deserts, which lie in coastal regions rendered arid by the presence of an offshore cold current, such as in Peru and northern Chile, demand separate consideration. Here, hydration made possible by condensation results in active disruption of the rocks although surface runoff remains practically negligible. In this case, therefore, the main agent of transportation is the wind. So powerful is it that it can move sand up mountain sides by as much as a thousand metres.

The distinction made between semi-arid and arid regions is the same as that made by many German writers, that is between *Randwuste* (marginal deserts) and *Kernwuste* (central deserts). This is a useful distinction in geomorphology because of the demonstrable difference in the rate of morphological development, the semi-arid regions being more dynamic than the neighbouring steppes or deserts. The infrequency of rain in the deserts is particularly favourable to the preservation of relic landforms of more humid conditions. Certainly some of the landforms in the central Sahara suggest formation by more vigorous morphological processes than operate at present.

In distinguishing, on the basis of the vegetation structure, between steppes, marginal deserts and central deserts, another factor must be considered, namely temperature variations. Of course, the action of wind and running water are affected by the degree of aridity, but these are processes of transportation rather than of disintegration and it is the latter which play the greatest role in the dry-zone morphogenic system.

The processes which result in the disintegration of earth materials dictate the volume of material removed from the exposed outcrops. Variation in temperatures above the freezing-point are responsible for the fragmentation processes found in all dry regions. Some workers consider that they act in a direct way through dilatation while others think that they act indirectly through desiccation which shatters the rocks. Whatever the truth of the matter, the sparse vegetation cover and the generally low humidities result in a wide temperature range at the ground surface. According to many authors, granite is thought to be particularly susceptible to this process which would explain the development of pediments and inselbergs in areas made up of this rock type. It might also explain why the crystalline rocks of the the Adrar area of Mauritania have been much more intensively eroded than have the sandstone plateaux. On other rock types, notably the sedimentary rocks and particularly shales, temperature changes above freezing-point do not appear to be capable of producing fragmentation: they are certainly much less effective than freeze–thaw action. For this reason, we distinguish between dry regions with severe winters and dry regions without a marked winter. Finally, the moisture régime of the rock surface itself should be considered. The action of condensation in the foggy deserts producing alternately wet and dry conditions which favour hydration has already been mentioned. For this reason, foggy deserts resemble deserts with cold winters in many ways.

The dry regions with severe winters experience severe fragmentation because snow, in contrast to rain, provides a constant supply of moisture at the ground surface and this helps frost-shattering. In Mongolia, Tibet and the Pamirs, the slopes are covered with vast screes and escarpment crests are serrated. Solifluction lobes may develop locally, although they are not active for more than a small fraction of the year. The slope forms suggest vigorous morphological change which is certainly not the case in deserts with no winter. In the deserts of middle-latitudes rock disintegration supplies an enormous volume of debris to the agents of transportation. This material becomes the load carried by periodic sheetwash and so aids the development of the piedmont plains. Except in the granite areas, the slow rate of disintegration in the tropical deserts inhibits relief development. Sandstone uplands seem to be evolving very slowly which would explain the preservation of old surfaces such as the backslope of the Grand Dhar which goes back at least to the Cretaceous.

Perhaps on the northern margins of the Sahara, the lower temperatures of the pluvial stages of the Pleistocene may have resulted in increased frost action. The superb piedmont surfaces which still survive in this region were formed at this time due to this greater rate of rock disintegration. Contrary to the opinion of Birot, they cannot be explained by the overloading of the present runoff waters because when the floods decline the waters are clear and tend to become incised. It is for this reason that piedmont slopes of Pleistocene age are more dissected in steppe regions with their more frequent floods than they are in truly arid regions where such floods are fewer. As the present rate of rock disintegration is about the same in both types of regions, floods in the drier areas transport the greater load.

Thus the morphological equilibrium of the Pleistocene piedmont surfaces is similar now to what it was then.

As in the middle-latitudes, so in the low latitudes the vegetation cover is typically of forest, at least where it is undisturbed. These two great forest zones of the earth come in contact in several areas, especially in south-east Asia. For this reason it is difficult to draw the boundaries between morphoclimatic zones in such regions.

Morphoclimatic boundaries are clearer in the northern part of Africa. The Sahara represents a very clearly defined zone separating wooded belts to the north and to the south. On its southern fringe, around the 17th parallel, the Sahelian zone represents the transition from savannas to desert. This is a grass steppe region with woodlands of acacia and baobab. Regular drought results in intermittent and poorly integrated drainage, sheetwash and rillwash being important on the impermeable rocks while the sandy members are practically areic. Wind is important only on alluvial deposits and in areas where the vegetation cover has been disturbed. Loamy soils are subjected to gullying which, because of the aridity, produces severe slopes. However, climates of the Pleistocene have endowed this area with a typically desert morphology containing a number of *ergs* and some fine inselbergs and *bajadas* substantially unaffected by current geomorphological processes. Despite the presence of woodland species, the Sahelian zone cannot be included in the same category as the savannas, for the rate of the soil-forming processes is slow, there are no aluminium and iron-rich soils and the dominant relief forms are those of the marginal deserts.

The Sahelian zone passes southward into the zone of the true savannas with their well-marked wet and dry seasons and rather higher annual precipitation (600–800 mm). Rain-forest begins to appear with an annual precipitation total of 1500 mm as long as the dry season is not too protracted. With such a protracted dry season the humid tropics, even given an annual precipitation greater than 1500 mm, are covered by other types of forest, such as those of Vietnam, the Indo-Gangetic plain or the Serra do Mar of Brazil. In the present state of knowledge, however, there seems to be no important structural difference between these two types.

The whole of the *humid tropical zone* enjoys high temperatures and freedom from frost, so that rock disintegration is limited. Rocky slopes evolve very slowly and fresh scree is not found at their foot. Any boulders that are found in such positions are usually coated with a weathering patina and well covered by vegetation, suggesting that they have been in position for a long time. They appear to be relics from the last dry climatic phase, at least in the Fouta Djallon, which would here correlate with the last glaciation. The slow rate of break-up of the rocks probably explains the persistence of precipitous slopes such as are found on sugarloaves, inselbergs and scarps. They must be virtually indestructible in the absence of frost action. Temperature variations, so important in the dry climates, are much less significant in the humid tropics especially under a forest cover. Humidities are higher, cloudiness is greater and the heating of rock surfaces during the daytime is less marked than in deserts. Moreover, the nights are not as cool; while temperatures may drop to 3° or 4°C in January in southern Mauritania they fall only to 12° or 15°C at Kindia in the Fouta Djallon. The fluted surfaces of sugarloaves and other monoliths are testimony of the insignificance of mechanical disintegration. These flutings are due to corrosion by meteoric waters flowing over the rock surfaces.

In contrast, however, the high temperatures and high rainfall figures result in intense chemical decomposition which is not matched elsewhere on earth. Despite the great volume of organic matter supplied by the luxurious vegetation it does not accumulate at the surface because humus decomposes so rapidly under these conditions that any well-drained soils may completely lack humus. The silicates, particularly the micas and the feldspars, are vigorously attacked and reduced to clays so that silica and alumina are liberated. The alumina and certain iron salts combine to form duricrusts or *cuirasses* sometimes as much as 10 metres thick. Even if this extreme development is absent, however, thick kaolinitic clays are produced. Silica is removed in pseudo-solution. The lime dissolves rapidly in the high rainfall. Almost all the primary rock minerals are attacked by chemical processes, the most resistant being quartz and some of the heavy minerals. It is for this reason that solid load in the form of sand and pebbles is so limited in the streams of this region. In the absence of frost action, the bedrock is less fissured and less eroded by streams with their smaller load than is the case in the middle-latitudes. Breaks of slope are not worn back and, as a result, the rivers of the humid tropics have characteristically stepped longitudinal profiles. Also, the stream channels themselves are very variable in their dimensions. Streams occupy shatter zones, clearing them of their loose debris without having the power to broaden them further by attack on the unweathered rock faces. Such features exist only as far as the savannas, disappearing in the Sahelian zone as the aridity reduces chemical weathering. Moreover, streams are fewer here and their mechanical activities are more marked.

The *humid tropical zone* consists of two major components, the savannas and the forests, each having its own morphogenic system.

The tropical savanna (including the tropical deciduous forests) where rainfall is lower than in the rain-forest proper and, moreover, has a very marked seasonal régime. The first falls of the wet season are often violent and fall on hard, parched ground. Because of the sparse vegetation cover they result in splash erosion and intense rillwash. The efficacy of the latter is limited to some extent by areas of dense grass cover but it can grow to take the form of sheetfloods. Areas of very low gradient are unevenly inundated and generally lack a co-ordinated system. Only a small part of this water finds its way into the stream channels which, in any case, cannot incise themselves as they lack the material to abrade their beds. This type of drainage, as shown by Dresch (1947) leads to colluviation. On topographical highs the water picks up soil fragments and dissolved matter which is then deposited in standing water in the depressions. This process, plus the lack of stream incision, is an important method of planation. It explains the great uniformity of almost horizontal surfaces over vast areas of the savannas.

Chemical weathering is less severe in the savannas than in the rain-forest because the amount of water which infiltrates is less. As the soils are less leached, ferruginous concretions are numerous. Cemented together they form *cuirasses*, and are more widespread here than in the forest lands although they are the result of allogenic ferric iron or paleoclimatic variations. Bauxitic or ferruginous *cuirasses* are the general rule on savannas in Brazil (*cangas*), Africa and south-east Asia. As they are so resistant to weathering, these *cuirasses* form protective surfaces, often making up low plateaux with abrupt edges undercut by steep-sided stream valleys.

Man has been an important influence on the evolution of the savannas, especially in Africa where man-made fires have been an important factor in the development of *cuirasses* and the gradual retreat of the forest margins. In other words, the savannas have been artificially enlarged. As the *cuirasses* are revealed by erosion, surface runoff increases but its effect is limited by the lack of any important physical weathering.

The tropical forest. The most important subaerial process here is chemical weathering which produces a thick regolith. Lime is readily dissolved and granites, gneisses and similar rocks reduced to clays so that they are covered almost everywhere by a friable weathering profile at least 10 m thick. The bulk of the rainwater infiltrates into this profile and so maintains the process of deep weathering. Where the granitic rocks are unjointed, they stand up in sharp contrast to this weathered layer in situations such as river banks, waterfalls and monolithic domes. As no weathered material is retained on their surfaces, these domes rise above the adjacent hill country which is continually attacked by chemical weathering beneath the forest cover. The slopes of these quite steep convex, forested hills, the *mar do mauros* or sea of hills, and the *meias laranjas*, or half oranges, of Brazil, are reduced by burrowing animals or falling trees which expose the ground surface to splash erosion or earth flow as, for example, in the Fouta Djallon. Springs carry great quantities of material in solution and may be the cause of slumps and even mud-bursts as, for example, in the forests of Colombia and Panama. Elsewhere, for example in French Guiana, the soil may be covered by a thick layer of litter which protects it so that runoff is not important. Elsewhere the decay of vegetable matter is so rapid that the soil surface is almost bare, causing some water to move as overland flow and carry suspended matter which is deposited at the slope foot. Some authors, such as Rougerie, accept a rather greater importance for overland flow under such conditions. However, this appears justifiable more on shales and mica-schists than on granitic rock and on semi-evergreen tropical forests rather than the evergreen type. The geomorphic importance of paleoclimatic changes appears to be important in the tropical forests as well as in the savannas, for relic Pleistocene landforms are quite common parts of Africa. For example, in the Ivory Coast pediplains of Sudanese type developed during dry phases of the Pleistocene right along the coast. Evidence suggesting formation of savanna plains around Kinshasa (formerly Leopoldville) has been reported. In Brazil, many monoliths have been stripped by landsliding during a period characterised by a tropical climate with a marked dry season which appears to correlate with the Flandrian transgression phase. On the edge of Lake Maracaibo in Venezuela climatic terraces made up of coalescent gravel fans are now covered by thick rain-forest. It follows that by no means all of the morphology of the humid tropical zone is significant in terms of present morphoclimatic activity. Only detailed work on current morphogenic processes in relation to the present landscape will succeed in defining the precise importance of the paleoforms.

REFERENCES

BÜDEL, J. (1948). Das System der Klimatischen Geomorphologie. *Verhandl. Deutscher Geographentag.* München, **27**, 65–100

DRESCH, J. (1947). Pénéplaines africaines. *Ann. de Géogr.*, **56**, 125–130

FOURNIER, F. (1960). *Climat et érosion.* Paris, P.U.F., p. 201

FRIEDLAND, V. N. (1951). Essai de division géographique des sols des systèmes montagneux de l'U.R.S.S. *Pédologie*, **9**, 521–53

KREBS, N. (1966). *Vergleichende Landerkunde* (3rd edn) Stuttgart, Kohler, 484 p. (pp. 65–77)

LAUTENSACHS, H. (1950). Granitische Abtragungsformen auf der iberischen Halbinsel und in Korea, ein Vergleich, *Petermanns Geogr. Mitt.*, 87–196

TRICART, J. (1958). Division morphoclimatique du Brésil atlantique central. *Rev. Géomorph. Dyn.*, **9**, 1–22

TROLL, C. (1944). Strukturboden, Solifluktion und Frostklimate der Erde. *Geol. Rundschau*, **34**, 545–694

15 Relationships Between Geomorphic Processes and Modern Climates as a Method in Paleoclimatology

LEE WILSON

THE distribution of landform types is related to factors such as the distribution of geologic structures and the distribution of geomorphic processes. The distribution of geomorphic processes is in turn largely controlled by the distribution of climatic types. Therefore, some aspects of the character of an equilibrium landform reflect the control of the dominant climatic type.

In this paper an attempt is made to relate climate types and the distribution of geomorphic processes, with some suggestion of resultant landscape characteristics. More importantly, the derived morphogenetic classification is correlated with actual climate types, using the genetic system of climatic classification developed by Strahler (1965). Strahler's system is based on the distribution of the earth's major air masses and seems to explain well the distribution of many natural phenomena, such as vegetation and soils (Oliver, 1969). The model presented here is in all ways to be considered as being qualitative, subjective, and (as in most models and classifications) artificial. It follows that there are assumptions and problems inherent in the following analysis. While most of these have implications too extensive for detailed discussion here, they are listed below; they should be borne in mind throughout the sequel.

(1) The analysis involves a very simplistic approach, and results in extremely general statements, mainly as a result of the following fact.

(2) The relationships between climate, process and landforms are not always well understood, especially in a quantitative sense; morphogenetic classifications will improve as these relationships become more precisely defined.

(3) The analysis largely ignores factors unrelated to climate. Relief, lithology, structure and time are assumed to be constant or unimportant (a patently false but unavoidable assumption). In general, the following analysis assumes a landscape of moderate to high relief,

with little variation in lithology and structure (for example, mountain areas underlain by crystalline rocks, and plateaux underlain by sedimentary rocks with constant structural trends). As areas considered become more complicated, the analysis becomes more subject to error; however the basic principles presented are presumed to be valid over most of the earth's land surface.

(4) It is assumed that equilibrium landforms can be distinguished from fossil landforms and, indeed, that parts of the earth's present surface are in equilibrium with today's climatic pattern.

(5) All natural phenomena resist classification; the distinct climatic and morphogenetic types here recognised merge into one another through transition zones which may be as broad and important as the discrete types.

(6) In the preliminary analysis primitive climatic parameters are used (mean annual temperature and precipitation); use of more sophisticated parameters is introduced later on in this chapter

(7) Finally, there is an abundance of terminology used in climatic geomorphology. Table 15.1 summarises the terminology used in this paper. The most important new concept is denoted by the term *climate–process system* (CPS), which is defined as sets of climatic conditions under which particular suites of geomorphic processes are dominant.

CLIMATE–PROCESS SYSTEMS

The first step in morphogenetic analysis is the establishment of specific climate–process relationships and the synthesis of these relationships into a classification of CPS.

Figure 15.1(a) indicates combinations of temperature and precipitation where, other things being equal, *mechanical weathering* is relatively enhanced or diminished. Mechanical weathering is at a maximum in regions having freeze–thaw and/or wet–dry cycles. In cold, moist regions where freeze–thaw processes are intense, mechanical disintegration is a dominant geomorphic process. Where freezing or thawing rarely occurs, or where moisture is absent, mechanical weathering is much less severe. Disintegration is also at a maximum in arid areas where desiccation processes (drying after wetting) are important. These processes include salt-crystal growth and wetting and drying of clay minerals.

Table 15.1 Definitions of terminology of morphogenetic classification

Term	Definition	Example: qualitative statements
Climate–process system (Wilson, 1968)	Concept relating climatic factors to geomorphic processes	Wind action is largely limited to dry climates. The arid climate–process system occurs where precipitation is low and evaporation is high. It is characterised by wind action, desiccation processes (such as salt-crystal growth), and very rare but intense rainstorms.
Morphogenetic system (used by Cotton, 1958)	Concept relating climate, process and landforms	Dry climates are typified by wind action which produces dunes and deflation basins. The arid morphogenetic system occurs where precipitation is low and evaporation is high. It is characterised by wind action, desiccation processes and infrequent rainstorms. The landscape is characterised by dunes, deflation basins, desert pavement, salt pans, rock shattered by salt crystals, cavernous weathering, *arroyos*, patterned ground (polygonal mud cracks, for example), and many other features.
Morphogenetic region (used by Peltier, 1950)	Actual area where landforms reflect present climate and processes	The Sahara Desert is an area where the arid morphogenetic system operates today. It has the climate and landscape characteristics of the arid system.
Paleomorphogenetic region (Wilson, 1968)	Area where landforms reflect past climate and processes	Much of the Congo Basin was an arid morphogenetic region during glacial maxima as is indicated by the existence of fossil dunes whose age may be radiometrically determined.

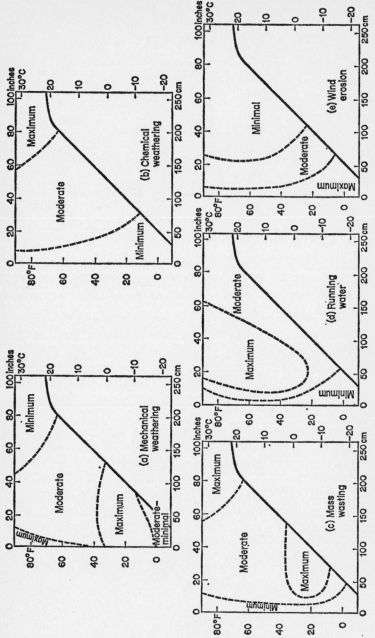

Fig. 15.1 Suggested relationships between various geomorphic processes and climatic parameters of mean annual precipitation and temperature

Mechanical weathering is shown to be at a minimum (though by no means negligible) in extremely cold regions, and in hot, wet climates. In the latter an abundance of organisms abets rock disintegration, but this factor is considered relatively minor. Figure 15.1(a) differs from figures produced by other writers (for example Peltier, Leopold *et al.*) mainly in the emphasis given to desiccation processes.

Figure 15.1(b) outlines the relative importance of chemical weathering under different climatic conditions. There is a multitude of chemical reactions and no one graph could summarise their relation to climate. Processes of solution, which are most important in monomineralic rock types, are excluded from discussion in order to make some first approximations. This permits Fig. 15.1(b) to be based on the fact that most weathering reactions involve water as an active or catalytic agent; hence chemical weathering becomes increasingly important as precipitation increases. Also, the rate of most chemical reactions increases with temperature, everything else being constant. Thus tropical climates promote the most intense chemical decomposition. The dense vegetation of tropical regions produces organic acids which further promote chemical decay.

In dry and cold regions chemical weathering is of moderate to minimum importance, although even in deserts and polar regions chemical processes may have considerable effect. If solution reactions are considered, Fig. 15.1(b) may still be broadly correct, since the total effect of solution increases with precipitation. Many chemical processes, particularly soil cycles involving solution and reprecipitation, are related to seasonal climatic cycles (see discussion of seasonality).

Figure 15.1(c) illustrates the probable importance of *mass movements* relative to variations in climate. Emphasis is placed on processes of solifluction and related phenomena in cold, wet regions, and on landsliding which is especially important in wet, hot (tropical) regions. These climates promote maximum weathering (in the former case mechanical, in the latter chemical) and hence promote accumulation of movable debris. Also, most mass movements are favoured by abundant moisture. Even in regions where mass movements are relatively less important (for instance, humid temperate regions), these processes are among the major landscaping agents. Mass wasting is favoured by seasonal climates, especially those with a pronounced wet season, or a season of convective precipitation.

Figure 15.1(d) outlines the variation of erosion by fluvial processes in terms of precipitation and temperature. As fluvial processes are the dominant agent sculpting most contemporary landscapes, this graph is critical to morphogenetic analysis. The figure differs considerably from those produced by other writers, and is based on modifications of a curve developed by Langbein and Schumm (1958). Both curves (Fig. 15.2) suggest that fluvial erosion intensities are not related to climate in a simple way. As precipitation increases there is

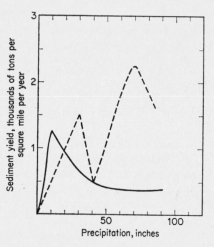

Fig. 15.2   *Sketch graph showing variation of sediment yield from drainage basins in relation to mean annual precipitation*

Solid line is sketched from Langbein and Schumm (1958). Dashed line is based on analysis of more extensive data than was available to Langbein and Schumm. Magnitudes of sediment yield are adjusted to basin areas of 259 sq km. Precipitation values are based on runoff records. Curves are based on data averaging, and scatter of raw data around curves is considerable

more water available for transporting sediment, hence erosion tends to increase with precipitation. However vegetal cover also increases with precipitation and tends to inhibit erosion. In semi-arid regions precipitation is sufficient to cause severe erosion, while vegetal cover is relatively scanty. Hence there is an erosion peak in semi-arid regions.

Under more humid conditions, vegetation inhibits erosion, despite the greater precipitation, while under more arid conditions precipitation is simply too scarce to promote maximum fluvial activity. The modified curve in Fig. 15.2 is based on data from over 1500 river basins and utilises approximately six times the number of data points available to Langbein and Schumm. The results available to date are consistent with the characteristics of the fluvial régimes listed in Table 15.3. Note that highest erosion rates occur under seasonal climates, especially where precipitation is seasonal. The equable régimes (equatorial, middle-latitude equable) show relatively low rates of erosion as do deserts. High rates of erosion under these régimes are generally traceable to environmental conditions other than climate, especially cultivation (Fournier, 1960; Douglas, 1967). Fluvial activity in polar regions is a function of glacial coverage and melting, and is at present largely indeterminate. The semi-arid peak for erosion intensity shown by Langbein and Schumm (at 25 to 38 cm mean annual precipitation) may actually occur under somewhat more humid conditions (63·5 to 89 cm mean annual precipitation); and erosion intensity is even greater in tropical regions, with maximum erosion peaking at approximately 190 cm mean annual precipitation.

Of course the modified curve of Fig. 15.2 cannot really be interpreted in terms of annual climatic parameters. Régimes which are semi-arid throughout the year are rare; rather, semi-arid conditions occur seasonally in combination with other climatic types. The tropical peak is probably most pronounced in monsoonal and savanna climates which have a definite dry period. Figure 15.1(d) is thus based on some consideration of seasonal effects. Maximum fluvial erosion is shown occurring under semi-arid conditions, with the understanding that such conditions nearly always occur in combination with other climatic types. In fact, it is preferable to state that maximum fluvial erosion is associated with *seasonal aridity*, rather than semi-aridity.

Figure 15.1(e) indicates that the effect of wind erosion is at a maximum in arid regions, where protection by vegetation is minimal. As precipitation and vegetation cover increase, the effect of wind action is sharply reduced. Cold regions are typically areas of strong winds, and limited vegetation: hence they also show considerable effects of wind erosion and deposition.

The relationships in Figure 15.1 are general, but I believe them to be consonant with what is presently known. The figure does not exhaust the realm of geomorphic processes, and additional diagrams could be drawn.

Based on Fig. 15.1 it is possible to delineate six CPS. These are shown in Fig. 15.3. The boundaries on Fig. 15.3 and the criteria on which they are based, are discussed in Wilson (1968). Table 15·2 summarises the CPS and tentatively converts them to morphogenetic

Table 15.2 Simple morphogenetic systems

System name	Equivalent Köppen climates		Dominant geomorphic processes	Landscape characteristics
Glacial	EF	Ice-cap	Glaciation Nivation	Glacial scour Alpine topography Moraines, kames, eskers
Periglacial	ET EM D-c	Tundra Humid microthermal	Wind action (freeze–thaw) Frost action Solifluction Running water	Patterned ground Solifluction slopes, lobes, terraces Outwash plains
Arid	BW	Desert	Desiccation Wind action Running water	Dunes, salt pans (*playas*) Deflation basins Cavernous weathering Angular slopes, *arroyos*

Semi-arid
(Subhumid)
	BS
	Cwa
		Steppe
		Tropical savanna
			Running water
			Weathering (especially
			  mechanical)
			Rapid mass movements
				Pediments, fans
				Angular slopes with coarse
				  debris
				Badlands

Humid temperate	Cf
	D-a
		Humid mesothermal	Running water
			Weathering (especially
			  chemical)
			Creep (and other mass
			  movements)
				Smooth slopes, soil covered
				Ridges and valleys
				Stream deposits extensive

Selva	Af Am	Tropical Monsoonal	Chemical weathering Mass movements Running water	Steep slopes, knife-edge ridges Deep soils (laterites included) Reefs

Table 15.3 Classification of climatic régimes (Strahler, 1965) showing relationships to air-mass dynamics, climate–process systems (CPS), and intensity of fluvial processes; an example of each type is given; combinations of régimes are important in many areas

Climatic régime	Air-mass dominance	CPS	Intensity of fluvial processes	Example
Polar	cA (year round)	Glacial; locally periglacial	Indeterminate	Greenland
Desert	cT (year round)	Arid	Low except intense erosion during storms	Sahara
Equatorial	mT (year round)	Selva	Moderate	Congo Basin
Middle-latitude equable	mP (year round)	Humid temperate; locally variable	Moderate-to-low	New Zealand Lowlands
Tropical wet-dry	cT (dry season) mT (wet season)	Seasonal oscillation from selva to arid or semi-arid	Highest known, especially where land use is intense	South-east Asia
Mediterranean	cT (dry season) mT (wet season) and/or mP (wet season)	Seasonal oscillation from humid temperate or periglacial to semi-arid or arid	Often very high	Coastal California
Continental	mT (warm season) cP (cold season)	Seasonal oscillation from glacial-periglacial to semi-arid-humid temperate	High-to-moderate	Northern Great Plains, USA

Air-mass symbols: c—continental; m—maritime; P—polar; A—arctic (similar to polar); T—tropical

systems by suggestions of associated landform characteristics. Figure 15.3 somewhat resembles a graph published by Tanner (1961) in which three extreme climates—glacial, arid, selva—dominate the landscape, with an intermediate zone comprising less extreme climatic types (in Fig. 15.3 these are semi-arid, humid temperate, periglacial). The three extreme climates represent three extremes of ground surface material: glacial–ice covered, arid–bare, and selva—completely vegetated. As Fig. 15.3 does not involve any

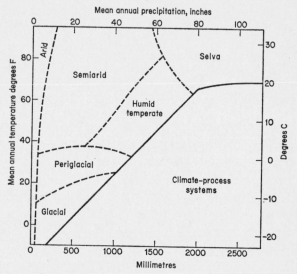

Fig. 15.3 Six possible climate–process systems (CPS) *each associated with a distinctive assemblage of landform characteristics*

Figure 15.4 shows the relationship between these CPS and actual climatic types (from Wilson, 1968). See text for importance of semi-arid belt (here probably exaggerated) to seasonal climates

consideration of seasonality, it does not represent the final step in the analysis.

SEASONALITY

The simplest way to introduce seasonality into morphogenetic analysis is not by studies of geomorphology but by study of distinctive climatic types. The climatic classification of Strahler (1965) as elaborated by Oliver (1969) is simple, based on natural climatic phenomena, and correlates well with distribution of the earth's physical characteristics. Seven basic climatic régimes are recognised (Table 15.3, Fig. 15.4).

Three of the basic climatic régimes correlate well with the three extreme climate–process and morphogenetic systems. Regions dominated by the equatorial *régime* can be expected to have landforms

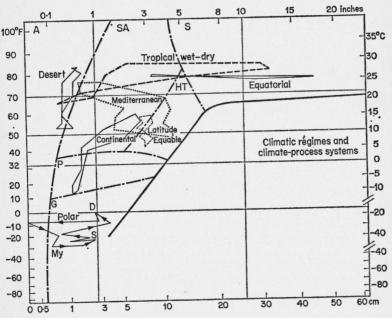

Fig. 15.4   Climatic régimes and climate–process systems

Graph shows monthly precipitation and temperature data as in a climograph. Scales were modified by Strahler (1965) to emphasise arid regions. System boundaries are given by dash-dot lines; systems indicated by letter abbreviations: A—arid, G—glacial etc. Polar, régime station (solid line) is McMurdo Sound, Antarctica. Equatorial—Padang, Sumatra (solid line). Desert—Willam Creek, Australia (solid line). Middle-latitude equable—Dunedin, New Zealand (dashed line). Continental—Winnipeg, Manitoba (solid line). Tropical wet-dry—Calcutta, India (dashed line). Mediterranean—Izmir, Turkey (short-dashed line). Régime diagrams courtesy John Oliver, Columbia University (from Wilson, 1968)

characteristic of the selva *system* (assuming, as usual, the absence of strong structural control, inherited landforms, and other complicating factors). Similarly the polar régime is approximately equivalent to the glacial system, although it also includes severely cold periglacial conditions. The desert régime is essentially the same as the arid system.

The middle-latitude equable régime of Strahler's classification corresponds to the humid-temperate CPS and morphogenetic system. However this régime and system is not extreme, and has probably been highly variable in the Quaternary. It is possible that few areas have been subjected to this climate and system long enough for equilibrium landforms to develop.

The remaining three régimes of Strahler's classification are strongly seasonal and involve more than one CPS. The tropical wet-dry and Mediterranean climates show precipitation seasonality. The tropical wet-dry régime involves annual oscillations between humid temperate and selva systems during the wet season to arid and semi-arid systems during the dry season. The Mediterranean régime is similar, although the seasonal oscillation is less, and selva conditions are rarely achieved. Areas subjected to these climates should show a combination of the landscape characteristics of the various morphogenetic systems, as well as features produced only under seasonal climates.

The continental régime involves seasonal temperature variations and produces an annual oscillation from glacial and/or periglacial conditions during the winter to semi-arid and humid temperate systems in the summer. Usually, there is also a seasonal shift in precipitation type, from winter frontal snowstorms to summer convective thunderstorms.

Table 15.3 and Fig. 15.4 represent the basic morphogenetic classification produced in this paper. The classification, while primitive, does have the merits of relative simplicity, and a high correspondence to natural conditions (especially existing climates). The primary drawback of the classification is that the morphogenetic aspect (that is, climatic influence on landform genesis) is as yet incompletely understood. The primary advantage of the classification is that it relates a simple scheme of process–landform relationships to a simple scheme of climatic classification. Moreover the climatic classification is based upon the dominant influence of various air masses, not upon rather specific criteria of temperature, precipitation, evapotranspiration or other climatic parameters. The classification is thus relatively easily translated into a tool for paleoclimatological analysis.

PALEOCLIMATOLOGY

Relict landscapes which can be identified in terms of one or more of the six CPS can in turn be related to one of the seven climatic régimes, and to the past distribution of air masses. This chapter does not attempt to establish precise criteria for landform–air-mass correlations, but merely suggests that such correlations exist and can probably be recognised.

The application of the morphogenetic approach may throw considerable light on the Quaternary climatic oscillations of Africa and South America. Application of the model to South America results in reinterpretation of morphogenetic data presented by Bigarella, Garner, and others (additional data and references to previous work can be found in Damuth and Fairbridge, 1970).

At the present time South America, east of the Andes, is dominated by mT or mE air masses in the Amazon basin, with zones of seasonal aridity toward both poles. The dry zones are due to continental or orographically modified mT air, which resembles cT air in ground effect. The climate régimes are selva and tropical wet-dry. Anomalous patterns, such as easternmost Brazil, will not be considered here.

There is compelling evidence, in relict landforms and inactive deposits, that past climatic conditions have differed from the present. Bigarella and his co-workers have identified a series of widespread erosion surfaces (pediments) which are being dissected by contemporary processes. Each surface is said to relate to a period of aridity, contemporary with glaciation elsewhere. Valley alluvium contains much coarse sediment which demonstrates a history of mechanical weathering in the source regions which today are mainly characterised by chemical weathering.

Arkosic sands of late Pleistocene age are found off the Amazon delta, and are taken as evidence of mechanical weathering conditions inland. Damuth and Fairbridge argue that weathering during Pleistocene arid periods produced the arkosic material which was flushed down large rivers during the climatic shift to humid conditions as the last glaciation terminated. A similar shift from an arid to a humid climate is held responsible for the anastomosing drainage pattern of the Rio Caroni and other streams, according to Garner. These streams reflect present-day modification of drainage produced by earlier arid conditions which were typified by braided streams and sheetwash on gravel-veneered surfaces.

Throughout the published observations on South American paleo-geomorphology the accent is on 'aridity', and analogies are sometimes drawn from the modern Peruvian desert. However the CPS model argues against quite such extreme conditions. Abundant sediments and widespread planation can only result from effective weathering and fluvial erosion. Neither is possible in truly arid climates. It is probable that much of the alluviation is due to flushing of sediment during the arid-humid transitions, as stated by Damuth and Fairbridge. However the origin of the sediment and the development of the surfaces are less well explained by processes acting over a short transition period.

Biological evidence, from pollen to iguanas, indicates that during glacial maxima the Amazon–Guiana region was dominated by savanna vegetation, with many forest refuges. The air-mass model supports this evidence since it is difficult to postulate a Quaternary paleoclimate which would exclude the seasonal influence of mT or mE air over much of South America. The probable paleoclimate is one of pronounced seasonality, with a definite wet season and an extreme dry season. Overall, the climate would be drier than at present, but not arid. The relict landforms can be explained by a savanna morphogenetic system, with mechanical weathering (for instance, desiccation), chemical weathering (for instance, along grain boundaries), and debris-laden floods at the onset of each wet season. True aridity would be limited to southern South America, or the periglacial regions, where relict dunes are found today.

The CPS model applies to other continents. The most common pattern is similar to South America—shifts toward more seasonal climates are likely to produce greater slope erosion, planation, and valley alluviation than shifts toward equable climates. Thus in Mediterranean Africa, glacial maxima correspond with periods of increased seasonal humidity (pluvials), with associated frost action. Erosion and planation are related to these periods, whereas in many areas today the climate is too dry for intense morphogenesis.

The analysis outlined above is not meant to be a definitive interpretation, but, rather, an attempt to show the potential and actual value of morphogenetic classification as a tool in paleoclimatological studies.

With reference to the general principle involved in such analyses, it should be noted that Douglas (1967) has raised the fundamental question of whether present-day fluvial erosion rates can be safely extrapolated into the past. It is probably true that absolute magnitudes of present-day rates are partly in response to human activity, and also that climatic conditions have varied considerably in the Quaternary. However, I believe that the relative judgements concerning fluvial erosion values (Table 15.3), if supported by analysis of modern records as presented in the first part of this paper, are probably valid indicators of erosion trends throughout the Quaternary. The data suggest that regions of high relief and weak rocks in Mediterranean climates should undergo denudation at a rate of 5 to 20 cm per thousand years. In areas such as California and Italy one can expect isostatic compensation of denudational unloading to result in perhaps 60 to 250 metres of uplift during the Quaternary, a factor which must be considered when interpreting raised shorelines.

Although climatic conditions have varied in terms of distribution, the basic climatic types of Fig. 15.4 have not altered drastically in the last few million years. Seasonal climates have always resulted in a seasonal lessening of the ground cover, and it seems likely that these régimes have always been characterised by high erosion rates. And, particularly in the case of the tropical wet-dry régime, it seems possible that human interference with the plant cover (for example, by burning) may extend far backward in time, perhaps to a time older than most fossil landscape features. In general, climatic shifts towards seasonal aridity probably result in accelerated upland erosion and stream aggradation, while shifts to equable climates result in upland stability and stream incision. The Langbein–Schumm concept, which considers vegetation–runoff relationships as critical in controlling erosion rates, is thus supported in modified form.

The morphogenetic model presented here considers contemporary geomorphic processes and equilibrium landscapes in terms of the present-day distribution of air masses and climatic régimes. It is considered that relict landforms can be related to this model and the pattern of air-mass dominance during the past established. Of the seven major climatic types dominating the present landscape, the three extreme climatic types and the three seasonal régimes are of particular interest; all have probably been of great importance throughout the Quaternary.

At this stage morphogenetic classification is qualitative, subjective, and general. Future research should improve this method of paleoclimatic analysis.

REFERENCES

COTTON, C. A. (1958). Alternating Pleistocene morphogenetic systems. *Geol. Mag.*, **95**, 125–36

DAMUTH, J., and FAIRBRIDGE, R. (1970). Equatorial Atlantic deep-sea arkosic sands and ice-age aridity in tropical South America. *Bull. Geol. Soc. America*, **81**, 189–206

DOUGLAS, I. (1967). Man, vegetation, and the sediment yields of rivers. *Nature*, **215**, 925–28

FOURNIER, F. (1960). *Climat et érosion*. Paris P.U.F.

LANGBEIN, W. B., and SCHUMM, S. A. (1958). Yield of sediment in relation to mean annual precipitation. *Trans. Amer. geophys. Un.*, **39**, pp. 1076–84

LEOPOLD, L., WOLMAN, M. G., and MILLER, J. (1964). *Fluvial Processes in Geomorphology*. W. H. Freeman, San Francisco, pp. 40–46

OLIVER, J. (1969). A genetic approach to climatic classification and regional climate analysis, *unpub. Ph.D. dissertation*. Columbia Univ., New York

PELTIER, L. C. (1950). The geographical cycle in periglacial regions. *Ann. Assoc. Amer. Geogr.*, **40**, 214–36

STRAHLER, A. N. (1965). *Introduction to Physical Geography*. Wiley, New York, pp. 111–16

TANNER, W. F. (1961). An alternate approach to morphogenetic climates. *Southeastern Geologist*, **2**, 251–57

WILSON, L. (1968). Morphogenetic classification. In *The Encyclopedia of Geomorphology* (R. W. Fairbridge, editor), Reinhold, New York, pp. 717–31

Index

Aar river 39, 68
Ablation 52, 53, 89, 239
  general (*see* Erosional lowering)
    zone 74
Abrasion 12, 76, 77, 78, 81, 85, 86, 136,
    233
Accordant junctions 37
Active layer 215, 216, 230
Adret 248
Africa 50, 64, 73, 155, 179, 180, 181,
    182, 217, 246, 250, 260, 264, 267, 281,
    282
Agassiz, L. 12
Aggradation 25, 39, 40, 41
Ahaggar region 180, 244
Ahlmann, H. W. 247
Aiguilles 80, 82
Alaska 64, 65, 142, 147, 187, 216, 231,
    254, 255
Albritton, C. C. 148, 151
Alexander, F. E. S. 174, 184
Algeria 62, 242
Allegheny, Mountains 62
    plateau 19
Alluvial, cones 73, 74
Alluvial fan 41
Alluvial plain 70
Alluvium 70, 112, 146, 150, 173, 259,
    260, 281
Alps 34, 69, 71, 78, 80, 81, 82, 84, 126,
    204, 210, 242, 246
      Apuane 255, 257
      Dinaric 126
      Franconian 223
Alsace 231
Alter, J. C. 139, 150
Altiplanation 142, 147
Amazon 219, 281, 282
Anatolia 118
Anderson, J. G. 142, 147, 150
Andes 161, 220, 244, 246, 248, 281
Angola 155, 157
Antarctic continent 71, 86, 87–90, 228,
    239
Antevs, E. 147, 150
Apennines 257
Appalachians 19, 67, 257
Arabia 188
Arctic regions 65, 250

Ardennes 157
Arête, (comb ridge) 69, 80, 81, 82, 84,
    86
Argentina 237
Arid regions 13, 41, 53, 54, 57–8, 78,
    136, 139, 161–3, 167, 228, 232, 239, 241,
    252, 261–4, 275
Arid regions (pseudo) 59
Arizona 43, 188
    river 62
Arkose 212, 215, 281
Armenia 118
Armorican Massif 155
Arrhenius, S. 134, 150
Asia 50, 61, 62, 63, 70, 96, 237, 250, 260,
    264
Asia Minor 207, 260
Atlas Mountains 118, 248, 260
Atmospheric composition 72
Atwood, W. W. 78, 90
Auelehmdecke 112
Aufeisboden 94
Australia 50, 207, 257, 258, 260, 261

Badlands 45, 62, 70, 74, 115, 163, 241,
    242
Bakker, J. P. 219, 225
Bajada 259, 264
Balance, between surface and linear
    erosion 121–2, 158, 160, 162, 164, 169
      between vegetation and mechanical
        processes 233–5
Barents Island 212
Barrancos 70
Basalt 97, 180, 212, 215, 242
Base-level 20, 24, 26, 33, 35, 47, 49, 50,
    102, 146, 154, 177, 180, 182, 184, 204–
    5, 206, 211, 247
    local 26, 42, 45, 46, 145, 146
Basin and Range province (U.S.A.) 13
Basins, aggraded 43
    arid 43–8
    deposits 44
    dissected 43
    glacial 39, 86
    internal, enclosed 162
Baulig, H. 178, 182, 184
Behre, C. H. 143, 150
Behrmann, W. 123, 129

Bergschrund 36, 77, 78
Bighorn Mountains (Wyoming) 78, 82, 84, 142
Birot, P. 15, 176, 184, 235, 263
Black Forest 211, 245
Blackwelder, E. 136, 150
Blockfields 64, 65
Blümcke, A. 58, 59
Blumenstock, D. I. 131, 133, 150
Bobek, H. 118, 129, 202, 225
Bolivia 220
Bolsons 46, 162
Borneo 188
Bornhardt 176, 189, 190, 191
Bornhardt, W. 49, 140, 150, 155, 169, 171, 184
Bourgogne 66
Brazil 67, 68, 96, 261, 264, 267, 281
Bremer, H. 209, 224, 225, 226
British Columbia 255, 256
Brittany 67, 157
Brückner, E. 58, 60
Bryan, K. 131, 133, 141, 144, 148, 150, 151
Büdel, J. 11, 12, 13, 16, 17, 104–30, 131, 133, 139, 151, 160, 161, 164, 168, 169, 171, 172, 173, 174, 175, 176, 177, 178, 179, 181, 182, 183, 184, 202, 209–27, 250, 254, 268

Caatinga 261
Cailleux, A. 11, 15, 16, 172, 176, 178, 179, 182, 185, 228–68
Cairnes, D. D. 142, 144, 146, 147, 151
California 68, 74, 245, 255, 258
Cambrian 67
Cameroons 96, 100, 172, 175, 177
Campos Cerrados 252
Canada 61, 188, 216, 230, 242, 254
Capillary action 57
Capps, S. R. 65, 143, 151
Carpathian Mountains 65, 142
Caucasus 71
Cévennes 245
Chalk 203, 207, 212
Chang, J. 187
Chaparral 259
Chebka 63, 70, 74
Chile 117, 244, 255, 256, 262
China 61, 70, 255, 259, 260
Cirque (*botner*, corrie, cwm, *karen*) 12, 13, 36, 39, 65, 76–90
  buried 86
  coalescent 80
  erosion 77–8
  headwall 76, 78, 81
  plan form 78–9
  Quaternary 160
  recession 76

Clay 49, 99, 100, 101, 116, 230
  minerals 217, 265, 270
  ironstone 212
Clayton, R. W. 172, 173, 176, 177, 184
Climate, accidents 179–81
  altitudinal zonation 104
  arid 15, 41, 49, 50, 52, 63, 141, 188, 198, 199, 200, 278, 282
  boreal 133
  change 15, 38, 50, 70, 71, 76, 107, 123, 124, 127, 131, 150, 157, 169, 172, 180, 221, 223, 225, 281
  classification of 14, 51, 92, 269
  continental 67
  cycle 273
  definitions of 51
  doldrum 94
  ecological 235
  equable 69, 112, 275, 282, 283
  equatorial, selva 67, 92, 171, 174
  glacial 69, 71, 141, 278
  humid 15, 50, 52, 61, 63, 141, 154, 188
  marine west-coast 133, 139
  Mediterranean 115, 117, 280, 283
  mesothermal (moderate) 198, 199, 200, 278
  moist desert 94
  monsoon 118, 183, 275
  morphological effects of 97–100
  nival 15, 52, 58–9
  oceanic 67
  periglacial 142, 278
  Pleistocene 124
  polar 55
  provinces 51–9
  Quaternary 181, 182
  regions 51
  savanna 118, 171, 217, 275
  seasonal 278–80
  seasonally wet-tropical 154–5, 171, 217, 218, 242, 252, 280
  semi-arid 188, 199, 200, 237, 278
  Tertiary 181
  tropical 198, 199, 200
  tundra 126, 161, 215, 224
  variations 91, 206–8, 281
  zones, belts 15, 51, 91–5, 104, 105–6, 124–5, 171, 202, 206, 207, 225
Climate–process systems (CPS) 14, 270–80, 281, 282
Climatic accidents 35
Climatic-geomorphological, morphoclimatic systems 11, 104–30
Climatic optimum 211
Climatology, air mass 14, 277, 280–81
Climograph 14, 279
Col 81, 84
Cold wave 230, 231, 248

olombia 261, 267
olorado, Plateau 183
  River 62
ongelifractate 145, 146
ongelifraction 143, 144, 145, 146
ongeliturbate 144, 146
ongeliturbation, geliturbation 138,
  139, 230, 231, 232, 254
onglomerate 67
ongo River 219
orbel, J. 203, 226
orrasion 136, 215, 235
otton, C. A. 11, 13, 16, 140, 151, 171–
  85, 189, 192, 271, 283
redner, W. 122, 129
reep 29, 94, 99, 138, 233, 234, 248, 256
retaceous 21, 50, 263
revasses 36, 77, 85
rimea 202
rustal Movement Theory 163–6
ryokarst 241
ryoplanation 142, 144, 145, 147, 148
uesta 191
uirasse 234, 265
ycle, arid 12, 13, 42–8, 50, 131, 140,
  180
  concept of 11
  cryoplanation 141
  geographical 12, 19–35, 50, 131,
    140, 141, 148, 153–4
  glacial 12, 13, 35–41, 76–90, 131,
    141, 149
  interruptions 32–5
  marine 137
  moderate 133, 140, 148, 149, 150
  normal 133, 140, 148
  of erosion 153–169
  periglacial 133, 141, 142–8
  pluviofluvial 141
  savanna 140, 171–2
  selva, hot humid 140
  semi-arid 140, 177
  soil 273
  stages of 13
Cyprus 193

Damuth, J. 281, 282, 283
Danube, gorge 210
  River 224
Dauerrumpf 164
Davis, W. M. 12, 13, 18, 19–50, 62, 64,
  105, 131, 133, 137, 140, 141, 143, 151,
  153–4, 159, 160, 161, 162, 163, 164, 166,
  169, 180, 205, 209, 226, 233, 235
Debenham, F. 89
Debris flow 116
Deccan 159, 217
Deflation 115, 119, 139, 180, 259, 260
Deforestation 117, 231, 259
Degradation 25, 39, 41

Dell, dellen 111, 113, 168, 211
Delta 24, 72, 281
Denudation 22, 106
  glacial 109
  postglacial 126
  potential 158
  rate 165, 168
Deposits, englacial 58
  glacial 149
  niveo-aeolian 238
Deserts 13, 63–4, 70, 74, 273, 275
  Australian 209
  central 15, 119, 121, 262
  classification of 119
  clay and silt 63
  cold 113, 263
  foggy 262, 263
  high-altitude 95, 120, 136, 139, 246,
    251
  low-latitude 136
  marginal 262, 264
  mid-latitude 139
  mountain 15
  peripheral 120, 121
  rock 63, 110, 230, 232, 240, 250,
    254
  sand 63
  trade-wind 50
  with severe winter 229
  without severe winter 229
Desquamation 67–8
Dew 44
Dilatation 262
Discordant junctions 37
Dissection 26, 34, 40, 179, 180, 181, 245
  257
Divides, consequent 23
  migration of 28, 29, 34
Dokuchaev, V. V. 235
Dolerite 218
Doline 116
Dollfus, O. 248
Dolomite 67, 207
Dominant processes 228–31
Double surfaces of weathering 174–5,
  218–19
Douglas, I. 275, 282, 284
Downwearing 182, 206
Drainage, areic 264
  areas 40
  basin 41
  coefficient 244
  consequent 42
  disintegration 48
  exoreic 242
  integrated 47, 48
  internal, endoreic 61, 71, 147, 242
  texture 193, 199
Drainage flushing 214–16
Dresch, J. 173, 174, 176, 184, 268

Dry boundary   53–4, 56
Dry season   15, 92, 93, 96–103
  morphologically significant   96–103
Dune sand   118
Dunes   43, 74, 192, 206, 259, 260, 282
  coastal   140
Duricrust   57, 234–5, 240–1, 242, 265
Dust, exportation   44, 47
  skin   119, 120
  storms   44, 46
Dynamic equilibrium   16

Eakin, H. M.   132, 144, 147, 151
Earth flow   94, 148, 234, 267
  subsurface   123
Ecotone   250
Enclave (geomorphic)   250
Endrumpf   153, 154, 156, 157, 163
Energy, budget   168–9
  kinetic   166
  potential   164, 166, 167
  solar   231
Entwicklung, aufsteigenden   160
  absteigenden   160
Epeirovariance   203–4, 217
Equatorial regions   172, 245
Equilibrium   91, 264
Equilibrium landforms   189–90, 192,
  269, 270, 280, 283
Equiplanation   142, 146, 147
Erg   180, 261, 262, 264
Erosion, accelerated   112
  agents and climate   68–72
  capacity, potential   110, 111, 112,
    122, 123, 124, 125–6, 127, 161, 182
  effectiveness   138, 139
  fluvial   15, 64, 69–70, 137, 163–6,
    169, 274, 282
  glacial   58, 69, 109
  intensity   235
  lateral   100–1, 110, 162, 163, 169,
    173, 176
  linear   13, 93, 106, 107, 121, 123,
    158, 216
  mechanical   111, 179, 260
  past conditions   70–2
  pluvial   135, 138
  rate   139, 158, 204, 210–13, 275,
    282–3
  remnants   13
  retrogressive or headward   30, 40,
    166, 167, 205
  sheet   233, 256, 257
  splash   257, 267
  vertical   100–1, 110, 123, 159, 166,
    167, 178, 181
Erosion surface   64, 92, 127, 128, 153–69
  dissected, destruction of   155–7,
    181, 225
  periglacial   147

Erosion surface—(*contd.*)
  solifluction   161, 163
  stepped   101, 102, 168
  Tertiary   157
Erosional lowering   175
Erzgebirge   157, 168, 211
Escarpment (scarp)   16, 43, 64, 66, 6
  73, 74, 98, 99, 100, 101, 122, 165, 17
  175, 176, 177, 263, 265
  retreat   176–8
  sapping   176
  structural   183
Etage, Etagement   229, 243–9
Ethiopia   220, 240, 244
Europe   63, 64, 71, 73, 78, 113, 127, 15
  182, 187, 198, 210, 219, 220
  central   65, 113, 116, 117, 128, 21
    213, 216, 222, 224, 250, 256, 25
    258
  Tertiary   182–3
Evaporation   52, 71, 112, 240
  potential   186–90
Exclave (geomorphic)   250
Exposure, aspect   247–8, 250
Extension phenomena   240–1

Facies (geomorphic)   73–5
  warm humid   73
Fairbridge, R.   281, 283
Falkland Islands   142, 147
Fanglomerate   118
Felsenmeere   65, 112
Felspar   98
Fennoscandia   113
Fiji Islands   178
Finland   112
Finsterwalder, R.   53, 58, 59
Firn   248
Fischer, H.   44, 56, 59
Flachmulden   158–60, 161, 162, 163, 165,
  168
Flächmuldenlandschaft   157
Flandrian transgression   267
Flood   43, 113
Flood plains   27, 28, 29, 94, 112, 116–17,
  128, 224
Flysch   207
Forest, Mediterranean   116, 117
  mid-latitude   111
  mist   93
  sclerophyll   94
  subpolar   94, 112
  subtropical   255
  tropical deciduous   252
  tropical, intertropical (selva)   122,
    140, 171, 229, 233, 252, 267–8
  tropical semi-deciduous   253
  tundra   126
Formkreisen   131
Fournier, F.   235, 268, 275, 284

rane 117
reeze-thaw 65, 107, 214, 228, 242, 246, 248, 263
  cycle 228–30, 253, 254, 270
reise, F. W. 140, 151
requency distribution 193, 194, 196–7
riedland, V. N. 246, 248, 249, 268
rost action 135–6, 138, 142, 146, 149, 160–1, 253, 254, 263, 265, 282
rost heaving 142, 148, 215
rost-shattering 58, 64, 139, 143, 148, 160, 228–31, 238, 252, 254, 263
  micro 230
  macro 230
Fuller, A. C. 187

Gaertner, H. R. von 217, 226
Garrigues 117, 231, 258, 259
Gaüfläche 183
Geosyncline 237
Germany 92, 207
Ghana 219
Gilbert, G. K. 25, 61, 62, 71, 78, 153
Glacial corrasion 36
Glacial deposition 12, 126
Glacial drainage 39, 40
Glacial erosion 12, 36, 38, 39, 126, 258
Glacial period 14, 65, 208
Glacial plucking 12, 36, 76, 81, 85, 86
Glacial regions 229
Glacial sapping 78, 79, 81, 82, 85, 86
Glacial scouring 36
Glacial stages in mountain sculpture 82–5
Glacial striation 253
Glacial theory 12
Glacial trough 36
Glaciation 16, 50, 124, 205
  advancing hemicycle 76, 80, 86, 87
  limits 58, 82
  receding hemicycle 76, 86, 87, 88
Glacier National Park 81, 82, 84, 85
Glaciers, adjustment to structure of 40
  alpine 41, 65
  Antarctic 87, 89
  branch 37
  cirque 65
  classification of 88–9
  consequent 40
  continental 76–7, 86, 89
  dendritic 88
  eroding 37
  expanded foot 88
  graded 37, 40
  horseshoe (cliff) 82, 88
  mature 35, 37
  mountain 76–7, 85–7, 88, 239, 240, 247, 253
  old 35
  piedmont 82, 88, 89

Glaciers—(*contd*).
  plateau 109
  polar 64, 247
  radiating 88
  recession 88
  subsequent 40
  temperate 247
  transporting 37
  trunk 37
  valley 12, 109
  youthful 35, 38
Glacifluvial deposits 253
Glacis 70
Gneiss 97, 155, 267
Goldthwait, J. W. 87
Gours 63
Grade 38, 39
  concept of 12, 16, 24–6, 30
Graded waste sheet 30, 31
Granite, granitic rocks 65, 68, 97, 99, 203, 262, 263, 267
Great Salt Lake 35, 71
Greenland 64, 69, 71, 144, 147, 161, 228, 230, 241, 253, 254
Groundwater 41, 55, 56, 57, 59, 144
  table 145, 174, 242
Gulf of Guinea 172, 179
Gully erosion 138, 140, 259
Gullying 139, 257, 258–9, 264
Gully-wash 110, 120

Hack, J T. 13
Hammada 63, 74
Hangrunsen 113
Harz Mountains 204, 220, 243
Hawaii 187, 188, 245
Hedin, S. 63
Heim, A. 61
Herbertson, A. J. 51, 59
Hettner, A. 133, 151, 153, 169
Himalayas 71, 220, 245
Hjulström, F. 221, 226
Hobbs, W. H. 12, 13, 76–90
Hobbs Glacier 87, 89
Hogback 191
Högbom, B. 142, 147, 151
Holocene 211, 212, 213, 224
Hoots, H. W. 139, 151
Horn (pyramidal peak) 81, 84, 86
Hult, R. 51, 59
Humidity 66–7, 71, 91, 122
  and relief evolution 72–3
Hydration 134, 262, 263

Ice, cap 85–7, 89, 228
  crust 238
  discharge 36
  extrusive 94
  ground 55, 94, 214–16, 251

Ice—(*contd.*)
　　needle 246
　　seasonal 216
　　segregated ground 215
　　sheets 41, 65, 71, 109, 125, 230, 236,
　　　239, 240, 241, 247, 253, 258
　　shelf 239
　　stream capture 40
　　streams 39–41
Ice Age 91, 124, 125, 128
Iceland 64, 126, 147, 230, 254
Incision (*see* Stream)
India 96, 155, 187, 188, 217
Infiltration 138, 259, 267
Inselberg 16, 48, 98, 101–3, 118, 120,
　　121, 122, 127, 155, 159, 161, 218–20,
　　259, 260, 261, 262, 264, 265
　　azonal 218
　　domed 176
　　landscape 118, 122, 140, 171–84
　　marginal 174–5
　　outlying 218
　　polycylic 176
　　shield 175, 218, 219, 223
Inselberglandschaft 171
Iran 70, 118
Isotropism 190–2
Ivory Coast 92, 236, 267

Jaeger, F. 15
Japan 187, 255, 257
Jaranoff, D. 118, 129
Jessen, O. 121, 129, 155, 159, 169, 181,
　　184
Johnson, D. W. 137, 148, 151, 173, 185
Johnson, W. D. 77, 78, 81, 90
Jotunheim 220
Judson, S. 143, 151
Jukes, J. B. 26
Jura 19, 72
Jurassic 67, 156

Kalahari 63, 71, 260
Karren 97, 98, 116
Karst 59, 72, 116, 117, 122, 203
　　Dinaric 202
　　dome 122
　　extratropical 203
　　landforms 117
　　Mediterranean 116
　　tropical 203
Kayser, K. 120, 122, 129, 218, 226
Kear, D. 178, 185
Kerben 112
Kerbspülung 116
Kerbtal 158–60, 161, 162, 165, 167
Kerguelen Islands 230
Kernwüste 262
Kesseli, J. 13
Kessler, P. 139, 151

Kettle hole 200
King, L. C. 173, 176, 177, 180, 18
　　185, 186, 192
Knetsch, G. 209, 226
Kohler, M. A. 187, 192
Köppen, W. 14, 51, 60, 107, 129, 13.
　　151, 198, 276
Körber, H. 223, 226
Korea 156, 188, 255, 257
Krebs, N. 121, 129, 155, 169, 255, 268
Krumbein, W. C. 196, 201
Kubiëna, W. L. 173, 180, 185, 221, 22
Kurowski, P. 53, 60

Labrador 64, 65, 255
Lakes 30, 34, 38, 39, 42, 43, 46
　　basins 26
　　consequent 26
　　glacial 58
　　Pleistocene 71–72
Landforms, and man 108–9, 117, 150
　　206, 224
　　and rock type 122, 158, 190–2
　　　241–2
　　characteristic distinctive 122, 123
　　　128, 133
　　classification of 19, 20
　　currently active 106–9
　　diagnostic 17
　　fossil 111, 113, 208, 209, 270
　　glacial 124, 186
　　normal 35
　　paleo 17, 91, 112, 123–5, 268
　　relief climax 221
　　remnants, relic 127, 128, 172, 181,
　　　183, 184, 221, 222, 241, 249, 251,
　　　252, 256, 259, 267, 280, 281, 282,
　　　283
　　selva 190
　　subaerial 17
　　submarine 105
　　temperate 186
　　tundra 186, 190
Landscape, apparently old 165
　　arid 13
　　belts 16, 91–5
　　discordant 71
　　facies 15
　　fluvial 158–60, 163
　　initial 23
　　mature 22, 33, 143, 153, 164, 182
　　Mediterranean 15
　　old 23, 32, 143, 153, 164
　　polycyclic 15
　　seasonally wet tropical 15
　　stepped 101
　　youthful 22, 23, 43, 153, 164
Landslide, landslip 22, 93, 94, 99, 116,
　　117, 138, 139, 200, 233, 256, 257, 267,
　　273

Langbein, W. B.   14, 274, 275, 283, 284
Languedoc   72, 231
Lapparent, A.   61
Laterite   93, 98, 99, 101, 218
Laurentian Highlands   19, 20, 41
Lautensachs, H.   203, 226, 255, 268
Lawson, A. C.   162, 169
Leaching   57, 115
Lehmann, H.   122, 129, 203, 226
Leopold, L. B.   14, 273, 284
Limestone   66, 67, 72, 84, 116, 242, 257
    crust   93, 94
Linsley, R. K.   187, 192
Loess   44, 47, 61, 70, 115, 146, 221, 224, 260
Lorraine   66, 67
Louis, H.   13, 109, 118, 129, 153–70, 181, 185
Lowdermilk, W. C.   149–51

McGee, W. J.   46
Machatschek, F.   53, 60
Mackin, H.   16
Madagascar   68
Magnitude and frequency   196
Main River   211, 222, 223, 224
Maquis   117, 258, 259
Mar do mauros   267
Marble   257
Margerie, E. de   62, 148, 152
Marl   207
    Keuper   212
Martonne, E. de   15, 61–75, 178, 184
Mass movement   13, 135, 138–9, 140, 141, 173, 242, 256, 273
Massif Central   126, 157, 204, 218
Matterhorn   81, 82
Matthes, F. E.   77, 78, 90, 142, 152
Mauritania   263, 265
Mauvaises terres (*see* Badlands)
Meanders   27–9, 34
Mechanical eluviation   214, 219
Meias larangas   267
Meinardus, W.   119, 129
Meltwater   65, 69, 112, 160–1, 253
Mensching, H.   112, 129, 224, 226
Mers de rochers   65
Merz, A.   54, 60
Mesa   67, 191
Mesozoic   49, 124
Mexico   62, 70, 261
Miller, J. P.   14, 284
Missouri   149, 150
Mittelgebirge   127, 168, 211, 219
Model, cyclic   12, 13
    Davis's   12
    regional circulation   14
Monadnock   48, 49, 183–4
Mongolia   263, 264
Monolith   190, 265, 267

Mont Blanc   69, 80, 245, 246
Monument   84
Moraines   69, 127, 253, 258
    englacial   253
    push   253
    supraglacial   230
    terminal, end   41, 82, 253
Morocco   44, 118, 248, 260
Morphoclimatic, analysis   123–9
    balance   112, 115
        and man   115
    classification   92–3, 106, 119, 250
    regions   13, 14, 16, 228–68
    zones   91–5, 109–23
Morphogenesis   92, 108
Morphogenetic climates   186–92
Morphogenetic regions   131–50
Morphogenetic, morphogenic   system   180, 228–68, 275–8, 279–80
Morphological hardness   203, 206, 207
Mortensen, H.   15, 106, 117, 119, 129, 209, 221, 226
Moselle River   210, 211
Mount Washington   87, 147
Mountain, regions   163, 205–6, 210, 229
    young-fold   122
Mud-bursts   267
Mudflow   138, 139

Nebelwald   93
    belt   94
    conditions   94–5
Néve   36, 40, 65, 77, 78
New Guinea   138, 188
New Zealand   178, 182, 210, 255, 256
Nickpoint   167
    migration   167–8
Niger   206, 219
Nile   206, 217, 219, 240
Nivation   142
Noë, E. de la   62, 148, 152
Normandy   34, 155
North America   71, 113, 156, 187, 210, 255, 258
Norway   80, 86, 256
Nunatak   86, 90, 230

Obst, E.   120, 122, 129
Oklahoma   188, 193, 194
'Old tropical earth'   208, 223–5
Oldland   163, 165
Oliver, J.   270, 278, 284
Outlier, 101, 175
Overland flow   234, 252, 256, 261, 267
Oxidation   134

Paleoclimate   14
    change   250, 267
    factors   249, 251, 252
    influence   258

Paleoclimatology 269, 280–3
Palsen 254
Panfan 162
Passarge, S. 11, 13, 16, 48, 49, 63, 64, 71, 91–5, 119, 129, 131, 133, 152, 153, 168, 169, 171, 185, 221, 226
Paulhus, J. L. H. 187, 192
Peat 230, 249
Pediment 162, 163, 175, 176, 177, 190, 191, 259, 260, 262, 281
  and esplanade 186, 191
  buried 173
  rock 122
Pedimentation 171, 175, 176, 177, 179
Pediplain 171, 173, 261, 267
Pediplanation 172, 173, 181, 182
Peltier, L. C. 11, 13, 14, 15, 131–52, 193–201, 273, 284
Penck, A. 14, 51–60, 105, 107, 109, 119, 129, 131, 133, 141, 152, 153, 164, 170, 209, 226
  climatic system 105–6
Penck, W. 153, 157, 160, 163, 164, 170, 204, 220, 226
Peneplain 13, 32, 34, 47, 50, 64, 72, 153–4, 160, 161, 162, 163, 165
  East Australian 183
  Tertiary 181–4
  uplifted 182
Peneplanation 173, 182
Penman, H. L. 187, 192
Pennsylvania 62, 149
Periglacial facies 65
  fossil features 127
  regions 17, 131, 229, 231, 242, 282
Permafrost (*tjäle*, perennially frozen ground) 112, 113, 148, 203, 208, 210, 213–14, 215, 216, 217, 219, 229, 231, 241, 242, 251, 254, 255
  extrazonal 241
  Quaternary, relic 229, 231, 252
Permeability 137, 199
Peru 94, 261, 262, 281
Petrovariance 202–3
Pfannen 98
Philippson, A. 117, 130
Phreatic regions 55–7
Phreatic zones 54
Piedmont 46, 263
  benchlands 177, 181, 184
Pipkrake 246
Planation 72, 204, 281, 282
  desert 181
  lateral 148
  savanna 171–84
Plants 41, 42, 43, 47
  cover 91, 92, 93, 108, 110, 117, 118, 119, 137, 138, 139, 146, 150, 199, 214, 230–34, 237, 238, 252, 254, 255, 259, 261, 262, 264, 274, 275

Plants—(*contd.*)
  cover effectiveness 201
  formations 207
Plateau 100, 102, 103, 180, 182, 183, 270
  relic, remnant 180
  structural 183
Playa 43, 46
Playfair, J. 31
Pleistocene 16, 50, 65, 82, 94, 128, 133, 139, 147, 211, 213, 220, 223, 224, 231, 252, 256, 258, 260, 263, 264, 267, 281
  features 125–8
  frost debris zone 126, 210
  processes 91
  tundra zone 126, 127, 210
Pluvial period 71, 92, 128, 180, 260, 263, 282
Polar regions 64–5, 74, 163, 167, 203, 209, 210, 217, 231, 273, 275
Polje 116
Pollen analysis 183
Polygonal ground 64, 74
Pore-water 242
Poser, H. 142, 144, 147, 152, 160, 170
Powell, J. W. 49, 62
Precipitation 14, 51, 54, 55, 78, 91, 110, 186–90
  seasonal 55, 92
  type 72
Preglacial surface 78, 80, 81, 84
Primärrumpf 153, 154, 157, 158, 160
Primärrumpfflächen 157
Probability 194
Process 19
  anthropogenic 258–9
  azonal 236–42
  biochemical 232–5, 256
  coastal 238–9
  endogenic 104, 202
  exogenic 104, 153, 168, 202, 204, 206, 217, 237
  extrazonal 236–42
  morphogenetic 107, 206, 208, 209, 232, 240, 244
  normal 20
  pedological 233–5
  physical, mechanical 232–5
  polyzonal, plurizonal 236
  zonal 236–42
Profile of equilibrium 25
Provence 62, 66
Province, subaerial 109–23
  subglacial 109
  submarine 109
Pyrénées 71, 126, 256

Quasi-equilibrium 49
Quaternary 35, 49, 63, 64, 65, 71, 111, 113, 161, 172, 178, 179, 180, 183, 280, 283

Racheln 163
Rain-forest 92, 93, 96, 171, 172, 234, 236, 242, 264, 267
Rainwash 148, 242
Rainy season 96–103
Randwüste 120, 262
Rapids 25, 26, 30, 210, 219
    ice 37, 38
Rathjens, C. 118, 130
Raynal, R. 248
Reiche, P. 139, 152
Regolith 55, 69, 92, 98, 173, 174, 176, 180, 214, 215, 234, 246, 253, 254, 256, 261, 267
Relief, analysis 222
    factor 198
    generation 221–5
        tectonic 225
    local 198
    micro- 231
Reliefsockel 157
Rhine, Falls 210
    Highlands 157
    plain 61
    River 205, 210, 223, 224
Richter, E. 77, 78
Richthofen, F. von 44, 61
Rill 63, 99, 116, 117, 242, 261
    erosion 138
    wash 119, 120, 121, 122, 137, 261, 264
Riss period 210
Roches moutonnées 253
Rock, aelotropic 190–1
    fall 234, 256
    fan 173
    glaciers 64, 65
    isotropic 190–1
    rivers 64
    steps 12, 101
Rocky Mountains 62, 64, 66, 68, 81, 85
Ross Island 89
Rougerie, G. 267
Runoff 54, 232, 234, 235, 248, 251, 254, 256, 257, 259, 260, 267, 283
    factor 54
    rate 117
Rumpfebenen 165
Rumpfflächen 155, 217–20
Rumpfstufe 165
Rumpftreppen 122, 157
Russell, I. C. 71
Russell, R. J. 143, 152

Sahara 44, 63, 69, 70, 71, 180, 209, 217, 244, 259, 260, 261, 262, 263, 264
Salina, salt flat 43, 46, 242
Salisbury, R. L. 25
Sampling, area 193–201
    corrections 196–8
    point 193

Sampling—*(contd.)*
    random 194–5, 197
    sequential 196
    systematic 194–6
    transect 193–4
Sand, drift 44
    exportation 44, 180
Sandstone 63, 65, 207, 212, 242, 263
Sapper, K. 123, 130, 139, 140, 152, 172, 185
Savanna 15, 74, 92, 94, 97, 120, 121, 157, 159, 161, 163, 189, 191, 202, 229, 234–5, 236, 252, 264, 266
    parkland 250
    regions 17, 155, 160, 163, 165
    short grass 250, 259–60
Scandinavia 41, 65, 142
Scarp (*see* Escarpment)
Scarp-foot depression 176–7
Schiefergebirge 211, 218
Schist 155, 207, 257, 267
Schmidt, W. F. 113, 130
Schotter 59, 100
Schumm, S. A. 14, 274, 275, 283, 284
Schunck, W. 53, 59
Scree 101, 143, 248, 268
Sea-level 13, 20, 21, 23, 47, 147, 154, 155, 156, 157, 159, 162, 163, 165, 174, 212, 217, 244, 245
    eustatic 156, 181, 184, 220
Sebkah 242
Sediments 84, 85
    yield 14, 274
Selling, O. H. 125, 130
Semi-arid region, province 56, 139, 171, 241, 252, 260–1, 274
Senstius, M. W. 188, 192
Shale 84, 263, 267
Sheetflood 46, 98–9, 118, 121
Sheetwash 98–9, 119, 120, 122, 139, 173, 177, 218, 260, 261, 263, 264, 281
Siberia 113, 126, 216, 230, 254, 257
Slate 207, 215
Slope, centripetal 42, 45, 49
    congeliturbate 144, 145
    consequent 23
    debris 92, 162
    denudation, erosion 116, 139, 142–7, 149, 150, 212–13, 215, 282
    evolution 245, 247, 256
    flattening 13, 22, 27, 48, 49, 145, 146, 148
    graded 25, 29–31, 33, 43, 46
    retreat 16, 103, 144–5, 147, 176
    shattering 110, 143
    superglacial 41
    talus 66
    weathering 101
Slope-wash 120, 137, 138, 140, 145, 148, 214

Smith, H. T. U.   139, 152
Snow, accumulation   69, 76, 77
   drifting   89, 232
Snowline   36, 52–3, 56, 104, 204, 209, 211, 246
Snow-melt   55, 58, 110, 116, 214, 216
Snow-patch   214, 248
Soergel, W.   148, 152
Soil   31, 32, 42, 47, 57, 67, 98, 106
   boreal   247
   (chernozem black earth)   111, 115, 260
   chestnut   111, 260
   erosion   115, 117, 150
   ferrallitic   260, 264
   ferric   260, 264
   forest   111, 247
   formation   106, 108
   glacial   126
   grey steppe   111
   leached   122
   local   93, 95
   mature   221
   moisture   93, 139, 234
   moor   93
   movement   110, 113
   orts   115, 120
   podsolised   111, 247, 260
   polygenetic   148, 219
   regions   136
   residual   73, 134, 137
   sierozem   120
   stripes   110, 214–15
   tropical   120
   tropical red earth   217
Sölch, J.   166, 170
Solifluction, gelifluction   92, 107, 109, 110, 112, 138, 139, 142, 147, 160, 214–16, 224, 232, 234, 242, 246, 254, 256, 258, 263, 273
   'deep'   242
   macro-   160–1
   rate   215
Solifluctionsrumpf   161
Solution   111, 116, 134, 233, 256, 273
Somaliland   180, 219
South Africa   49, 63, 122, 187
South America   14, 250, 255, 281
Spitzbergen   64, 65, 127, 142, 144, 147, 161, 211, 212, 216, 222, 228
Spreitzer, H.   164, 170
Steche, H.   136, 139, 152
Steppe   70, 71, 74, 99, 209, 231, 235, 252, 258–60, 262, 263
   degraded   229
   desert   111, 113, 120–1, 258
   dwarf bush   94
   extratropical   111, 113
   feather grass   113, 115

Steppe—(*contd.*)
   forest   115, 126
   gorges   113, 116
   high   94
   loess   126, 127, 210
   Mediterranean   118
   semi-desert   229
Steppenschluchten   113
Strahler, A. N.   221, 226, 269, 277, 278, 280, 284,
Stream, ability   25, 26, 164
   action   137–8, 149, 153
   adjustment to structure of   33, 46
   antecedent   34, 42
   braided   117, 173, 216, 281
   channels   39, 54, 116, 117, 159
   consequent   23–7, 45
   density   99, 244, 257
   disappearing   59
   discharge   220, 240, 244, 245, 256
   diversion   40
   equilibrium   158, 164–5, 167
   graded   25–8, 30–1, 33, 178
   gradients   24, 154, 159, 161, 163, 164, 169, 182, 210, 217, 219
   incision   16, 33, 40, 173, 182, 212, 219, 234, 247, 256, 260, 283
   insequent   26
   load   24, 26, 28, 159, 160, 162, 163, 164, 166, 167, 169, 173, 182, 216, 220, 234, 240, 256, 257, 263, 265, 267
   mature   25, 28
   old   25, 28
   profile   162, 165, 210, 219, 220, 221, 265
   seasonal   55, 99–100, 117, 162, 173, 215–16, 219, 244
   stage   56, 57, 70, 112, 116, 117, 215
   subsequent   26, 45, 46
   torrential   244–5, 247, 257
   underground   100, 203
   youthful   25, 28
Stream systems   46, 101, 116, 224, 244, 257, 260
   centripetal   42–3
   relic   260
Structure   19, 20, 32, 35, 38, 40, 46, 67
   and glaciation   84–5
   and landforms   119, 120, 183, 191
Subarctic regions   136, 147
Subsidence, depression   34, 148, 203–4, 206, 237
Subsidiary processes   228–31
Sudan   68, 173, 217, 222
Sugarloaf   68, 172, 176, 190, 265
Supan, A.   51, 60
Surface, basal   174
   'biscuit-cut'   86
   planation   13, 168

Surface—(*contd.*)
    preglacial 81
    roughness 200, 238
    uplifted 220
    wash 116, 223
Sweden 54, 112, 147
Syenite 97, 98

Taber, S. 215, 227
Taiga 230, 241, 248
Talus 62, 66, 143, 144
Talweg 206, 210
Tanner, W. F. 14, 186–92, 278, 284
Tarr, R. S. 87
Tasmania 255, 256
Taylor, G. 89
Tectogenesis 119
Temperate regions 65–6, 67
Temperature, range 97, 243
    rock surface 243
    Tertiary European 183
Terra, H. de 142, 147, 152
Terraces, climatic 267
    periglacial 148
    rock 85
    solifluction 110–11, 246
    stream 148, 161, 210, 224
    'transition' 223
    Würm 224
Terrain, analysis 193–201
    parameters 198–201
Tertiary period 21, 49, 107, 124, 127, 128, 173, 178, 225
Thorbecke, F. 15, 96–103, 131, 133, 152, 172, 185
Thornthwaite, C. W. 131, 133, 150, 250
Thunderstorm 121, 257, 280
Tibet 120, 161, 263
Tienhaus, R. 217, 227
Topography, composite 34
    knife-edge 186
    monogenetic 172
    morainal 200
    polygenetic 133, 137, 148–50, 172
Tor 189, 190, 191
Transportation agents 137–40
Treeline 56, 160, 161, 190, 204, 209
Tricart, J. 11, 15, 16, 172, 176, 178, 179, 182, 185, 221, 227, 228–68
Trockenshuttzone 111, 118
Troll, C. 107, 120, 130, 133, 139, 141, 152, 160, 161, 170, 186, 192, 236, 247, 254, 268
Tundra 92, 94, 230, 232, 240, 248, 254
    montane 247

Ubac 248
Uinta range 78, 84
Ukraine 116, 259
Uniformitarian approach 172

United States of America 50, 61, 66, 71, 73, 78, 162, 187, 191, 198, 237, 250, 259, 260
Uplands, fretted 79, 80–1, 84, 86
    grooved 79, 80, 82, 86
    monumented 81–2, 84
Uplift, 19, 23, 32, 33, 34, 49, 70, 72, 156, 166, 167, 175, 178, 179, 180, 181, 184
    isostatic 211–12, 237, 283
Uvala 116

Valleys, alpine 39, 68
    anastomosing 63, 70
    asymmetrical 27, 100
    blind 63
    broken-bedded 38
    consequent 23, 177
    corrasion 111, 113
    deepening 216
    drowned 34
    floor 210, 212, 216, 220, 224, 234
    formation of 210–13, 216
    glacial 36, 79, 81
    graded 182
    graded side slopes 29–31
    hanging 37
    incised 67, 68, 162
    insequent 177
    lowering 212
    mature 24, 148
    network 100
    profile 212
    Quaternary, Pleistocene 112, 183, 211, 224
    steps 212
    trunk 70
    U-shaped 80, 86
    V-shaped 27, 112, 158
Vegetation, man-modified 231, 233, 257, 258, 283
    natural 111, 260
Venezuela 261, 267
Villafranchian 233, 224

Wadi 55, 71, 74, 202, 206, 261
Waibel, L. 121, 130
Waldsteppe 113
Wallén, A. 54
Walther, J. 63
Ward, R. de C. 51, 60
Waterfalls 25, 26, 30, 33, 34, 210, 212, 219, 220, 267
Water-holding capacity 234
Wave action 137, 238–9, 253
Weathering, chemical 20, 41, 66, 67, 73, 98, 118, 122, 134–6, 139, 163, 172, 177, 178, 182, 217, 219, 234, 238, 252, 255, 256, 257, 265, 266, 273, 281, 282
    deep 67, 93, 99, 123, 173, 203, 209, 217, 255, 267

Weathering—(*contd.*)
directional 14
front 218
honeycomb 63
insolation 119, 120
mantle 66, 73
mechanical, physical 20, 41, 55, 57,
66, 69, 73, 74, 77, 85–6, 96, 97–8,
118, 134, 136–7, 139, 145, 146, 162,
216, 238, 255, 257, 261, 265, 270,
273, 281, 282
peneplanation by 142, 147
profile 217
rate 137, 139, 163
regions 135, 136
relic 174
salt 119, 120
subglacial 58
Wentworth, C. K. 140, 152, 187
Wilhelmy, H. 209, 221, 227
Willis, B. 70, 175, 176, 185
Wilson, L. 11, 14, 269–84
Wind action 43, 44, 47, 69, 70, 126, 135,
139–40, 146, 232, 233, 237–8, 248, 253,
259, 260, 261, 262, 275
Wirthmann, A. 205, 212, 225, 227
Wisconsin period 149
Wissman, H. von 162, 169, 170
Woeikof, A. 52, 60
Woldstedt, P. 183, 185
Wolman, M. G. 14, 284
Wood, B. L. 178, 185
Wright, F. E. 89, 147, 152
Würm period 128, 210, 211, 224

Yaalon, D. 187
Yardang 63
Yellowstone National Park 82, 83
Yukon 216, 255

Zeuner, F. E. 146, 152
Zonal concept 16, 235–6, 249
Zonation, latitudinal 243–9
vertical 243–9
Zone, arid 58, 163, 209, 232
bioclimatic 235
boreal conifer 112, 231, 232, 235,
250, 251
circumpolar forest 110
continental 108, 256–7
deciduous forest 112
dry debris 108, 111, 115, 116, 117,
118–21, 122, 125, 128
equatorial 93, 96, 176, 179, 251
extratropical desert 108, 113
extratropical valley formation 207,
208

Zone—(*contd.*)
frost debris, frost rubble 108,
109–10, 111, 113, 116, 117, 120,
122, 125, 126, 127, 209–10, 215–16,
251
glacier, glacial 110, 122, 125, 127,
207, 208, 209, 239, 240, 251, 253
high-altitude (cool) desert 108
humid mid-latitude 155–7, 209, 251
inner tropical with mature soils
108, 121, 122–3, 125, 128
marginal desert 108
maximum precipitation 244, 246
Mediterranean 108, 115–18, 121,
123, 125, 128, 229, 251, 252, 257–8
mid-latitude forest, with mild winter
229, 252
with severe winter 229, 252
mid-latitude (mixed) forest 112,
229, 231, 234, 251–2, 254–8, 259
morphoclimatic, climatic–morphological, climatomorphological
105, 108–23, 208–20, 236, 237,
249–68
oceanic 108, 256
of aeration 174
of frost heaving 215–16
periglacial 251, 253–4
permafrost (*tjäle*) 108
pronounced valley formation 207,
208, 209, 216, 225
Quaternary, morphological and climatic 180
savanna 15, 96–103, 162, 168
sheetwash 108, 110, 121–2, 123,
125, 217, 251
sheetwash subtropical 108
sheetwash tropical 108
soil flow 109–10, 117, 121
solifluction 160–1
steppe 108, 115, 232, 251
subpolar 94, 108, 112
subtropical pediment and valley
formation 207, 208
temperate 16, 62, 71, 163, 231
temperate mature soil (*ortsboden*)
108, 111–15, 116, 117, 120, 121,
122, 123, 125, 126, 127, 128, 129,
251
Tethyan 236
transition 95, 96
tropical 71, 217, 252–3
tropical hot desert 108
tropical planation surface formation
207, 208, 209, 216–20, 223
tundra 108, 110–11, 112, 125, 126,
127, 209, 215–16, 251, 255